Advancing Global Bioethics

Volume 10

The book series Global Bioethics provides a forum for normative analysis of a vast range of important new issues in bioethics from a truly global perspective and with a cross-cultural approach. The issues covered by the series include among other things sponsorship of research and education, scientific misconduct and research integrity, exploitation of research participants in resource-poor settings, brain drain and migration of healthcare workers, organ trafficking and transplant tourism, indigenous medicine, biodiversity, commodification of human tissue, benefit sharing, bio-industry and food, malnutrition and hunger, human rights, and climate change.

More information about this series at http://www.springer.com/series/10420

Henk ten Have
Editor

Global Education in Bioethics

Springer

Editor
Henk ten Have
Center for Health Care Ethics
Duquesne University
Pittsburgh, PA, USA

ISSN 2212-652X ISSN 2212-6538 (electronic)
Advancing Global Bioethics
ISBN 978-3-030-07695-5 ISBN 978-3-319-78984-2 (eBook)
https://doi.org/10.1007/978-3-319-78984-2

Printed on acid-free paper

This Springer imprint is published by the registered company Springer International Publishing AG part of Springer Nature.
The registered company address is: Gewerbestrasse 11, 6330 Cham, Switzerland

Preface

The Third International Conference on Education and Ethics (in Curitiba, Brazil, 2015), organized by the International Association of Education in Ethics (IAEE) highlighted important dimensions of ethics teaching. Several contributions particularly emphasized the need for global perspectives and approaches in global ethics education. It can be argued that with the advancement and expansion of mainstream bioethics into global bioethics, there is a concomitant need to expand and broaden ethics education. The current approaches that are primarily focused on individual health practitioners as well as professional associations should be amended with approaches focused on global perspectives, so that ethics education will acquire a real global dimension. This will imply, for example, examining global rather than individual problems in clinical medicine and research. It will also entail the assumption of a global framework of ethical principles and values. However, many theoretical and practical challenges exist. One challenge concerns the goals of global ethics education. Another has to do with the conditions of possibility for ethics education at global level. Furthermore, there are practical challenges since examples and best practices of global education in the area of bioethics are relatively scarce. Some of the presentations in Curitiba addressed these issues; they have been elaborated, renewed, and revised as chapters for this book.

The book aims to address the above challenges and to provide in-depth analyses of how they can be overcome. It will have three parts. The first is addressing the theoretical background of globalization and its implications for ethics education. The second part examines the goals of global ethics education, as well as the challenges that are presented in various cultural, social and political contexts, and economic inequalities. The third part presents and analyzes various examples, methods, and practices of global ethics education.

Center for Healthcare Ethics
Duquesne University
Pittsburgh, PA, USA
March 2018

Henk ten Have

Contents

About the Authors

Ayesha Ahmad is a philosopher with a PhD in medical ethics who is working in global health. She is a lecturer at St Georges University of London in Medical Ethics and Law, and an honorary lecturer at the Institute for Global Health, University College London. She specializes in mental health and the humanities in the context of gender-based violence in conflict developing trauma therapeutic interventions using traditional storytelling in extreme situations such as Afghanistan. Her interest in religion and culture in medicine stems from being a member of the Clinical Ethics Committee at Great Ormond Street Hospital for children in London. She is also a writer of poetry.

Solomon Benatar is Emeritus Professor of Medicine, University of Cape Town, South Africa, and Distinguished Senior Scholar, Dalla Lana School of Public Health, University of Toronto, Canada. Recent publications include *Global Health and Global Health Ethics* (Benatar S & Brock G. Eds. Cambridge University Press 2011) and *Global Health Governance and Global Power: A Critical Commentary on the Lancet University of Oslo Commission Report* (Gill S & Benatar S R. International Journal of Health Services. 2016; 46 (2): 346–65).

Leonardo D. de Castro is Professor in the Department of Philosophy at the University of the Philippines, Diliman. He has been involved in bioethics education for healthcare professionals and teachers, ethics committee members, and secondary school children. As the Vice-Chair of the UNESCO International Bioethics Committee, he participated in the development of the Bioethics Core Curriculum used by the UN Agency in its Ethics Teacher Training Program.

Volnei Garrafa is Director of the UNESCO Cathedra and Post-Graduate Program in Bioethics – University of Brasília (Brazil), Member of the International Bioethics Committee of Unesco (2010–2017), and Editor Chief of the Brazilian Journal of Bioethics. He was co-founder and President of the Brazilian Society of Bioethics (2001–2005) and Latin American and Caribbean Network of Bioethics – Redbioética/UNESCO (2003–2010).

Willem A. Hoffmann is Professor in Biomedical Sciences and the current Chairperson of the Research Ethics Committee at the Tshwane University of Technology in Pretoria, South Africa. In 2012, he completed the UNESCO Ethics Teacher Trainer Course in Dubrovnik, Croatia. His latest publications include two chapters in the *Encyclopedia of Global Bioethics* (Springer 2016). He is a co-editor of *Ethics for Healthcare Professionals – African and South African Perspectives* (Springer, in preparation).

Terry Maksymowych has been a bioethics educator at the Academy of Notre Dame de Namur in Villanova, USA, for over 30 years and is also an adjunct professor at Villanova University, teaching clinical ethics to healthcare professionals. She holds graduate degrees in biology, theology, and philosophy and holds a PhD in philosophy (clinical ethics) from Villanova University. She writes articles, offers teacher inservice programs, and presents at national and international conferences about teaching bioethics at the middle/high school level.

Flávio R. L. Paranhos is a vitreoretinal surgeon; Professor of bioethics at the Pontifical Catholic University of Goiás in Goiânia, Brazil; and a member of the Brazilian National Commission for Research Ethics (CONEP), Brasília, Brazil. He holds a PhD in Ophthalmology, a master degree in Philosophy, and a PhD in Bioethics. He was a research fellow at Harvard Medical School (1994–1995). His latest publications include a chapter in the *Encyclopedia of Global Bioethics* (Springer 2016) and the article 'Bioethics of Intervention and the Case of Drugs Bevacizumab and Ranibizumab for Retinal Diseases' (*Revista Iberoamericana de Bioética*).

Renzo Pegoraro is Chancellor of the Pontifical Academy for Life, Vatican City. He was Director of the Fondazione Lanza in Padua (Italy), 1994–2014. His latest publications are: I. Carrasco de Paula, V. Paglia, R. Pegoraro (eds.), *Virtues in the Ethics of Life* (Pontifical Academy for Life 2017); L. Caenazzo, L. Mariani, R. Pegoraro, *Converge of New Emerging Technologies: Ethical challenges and new responsibilities* (Piccin 2017); R. Pegoraro and C. Petrini, *Committees: Clinical Ethics Committees*, in H. ten Have (ed.) *Encyclopedia of Global Bioethics* (Springer 2016).

Thiago Rocha da Cunha is Professor of Post-Graduate Program in Bioethics at the Pontifical Catholic University of Parana in Brazil, and researcher at the Center for Bioethics and Diplomacy Studies in Health in Fiocruz, Brazil. He also is a member of the Latin American and Caribbean Network of Bioethics, Redbiotica/UNESCO, and Director of the Brazilian Society of Bioethics (since 2014).

Jan Helge Solbakk MD, Theologian, and PhD (ancient Greek philosophy), is Professor in Medical Ethics and Head of Research at the Centre for Medical Ethics at the University of Oslo, Norway. Recent publications include Manchola Castillo C, Solbakk JH (2017). 'Bioethics and imagination: towards a narrative bioethics committed to social action and justice.' *Med Humanit*ies 2017;1:1–6, and Santos M, Solbakk JH and Garrafa V. 'When hard evidence does not mitigate the rise of reimbursement-based medicine: The case of bone metastasis radiation treatment.' *Journal of Medical Ethic*: dx.doi.org/10.1136/medethics-2016-103607. At present, he is writing a book on *Bioethics and Human Development* (Routledge).

Henk ten Have has been Director of the Center for Healthcare Ethics at Duquesne University in Pittsburgh, USA (2010–2018). He was former Director of the Division of Ethics of Science and Technology in UNESCO, Paris (2003–2010). His latest publications are *Global Bioethics: An Introduction* (Routledge 2016), *Vulnerability – Challenging Bioethics* (Routledge 2016), and *Encyclopedia of Global Bioethics* (Springer 2016).

Isidro Manuel C. Valero is Assistant Professor of Philosophy at the University of the Philippines, Diliman. Specializing in Philosophy for Children, he has undergone training at the Institute for the Advancement of Philosophy for Children (IAPC) based in Montclair University, New Jersey. He has conducted seminar-workshops on Philosophy for Children for public school teachers in Manila.

Aimee Zellers is an Assistant Professor of Philosophy at Carlow University in Pittsburgh, USA. She recently served a 3-year term as the Michele R. Atkins Endowed Chair for Ethics Across the Curriculum (2013–2016) at Carlow University. In this capacity, she facilitated the development and implementation of a university-wide ethics curriculum.

Introduction

Today, globalization is more criticized than ever before. Anti-globalism seems to be the proper attitude nowadays, especially in Western countries that have imposed neoliberal globalization upon the world and have benefitted most from it. Now they blame globalization for global problems of unemployment, immigration, and refugees. It is also obvious that globalization is associated with increasing inequality. Only an elite minority is benefitting from global trade while the majority of the world population have not seen improvement in their life and work conditions. However, in many of the critical discourses, the underlying roots of global problems are not really addressed. It is not so much globalization that is problematic but its foundational value framework that is determined by neoliberal ideology. This framework offers a "neoliberal fantasy" imagining everybody as an entrepreneur, making oneself into a successful and wealthy global citizen. In practice, there is a long narrative of humiliation, disrespect, envy, rivalry, resentment, and failure. People feel abandoned by their governments and politicians. Globalization preaches formal equality while it is associated with enormous differences in power, ownership, status, and health. It imposes an impersonal economic order that primarily benefits a fortunate elite minority. For many, global processes only produce a "sense of being humiliated by arrogant and deceptive elites…" (Mishra 2017).

Neoliberalism assumes specific values; the primacy of self-interest, competition, and private ownership. It is driven by a specific philosophy of individualism. It assumes that individuals are free and responsible for their own well-being. An unrestricted market will offer them all opportunities for development and advancement. Social safety networks and protection are unnecessary and counterproductive (Ten Have 2016). Market ideology, in the words of Mishra, "offers a dream of individual empowerment to all" and has therefore been extremely attractive in the modern world (Mishra 2017, p. 324).

Against this backdrop, global bioethics is emerging as a new moral discourse. It articulates other values that can reconstruct and redirect the processes of globalization. Values such as solidarity, justice, vulnerability, and protection of biodiversity. It demands attention to the social and environmental context in which individual human beings can flourish. Global bioethics therefore goes beyond the individual perspective of mainstream bioethics that prioritizes the ethical principle of respect for autonomy, and that regards ethics primarily as a matter of personal decision-making rather than as a social intellectual and practical endeavor.

The focus of this book is on global education in bioethics. It highlights the question: what are the implications of globalization for the teaching of bioethics? This question is also affected by the neoliberal framework in specific ways. Political leaders nowadays not only ridicule globalization but also education. Policy-makers promote the simplistic view that education only serves to obtain jobs. Ages of civilization, however, illustrate that the purpose of education is not, or at least not primarily, economic. We do not have education to create jobs or prepare young people for jobs. The ultimate aim of education is transformative. In this context, reference is often made to Jean-Jacques Rousseau: "We are born weak, we need strength; we are born totally unprovided, we need aid; we are born stupid, we need judgment. Everything we do not have at our birth and which we need when we are grown is given to us by education" (Rousseau 1979, p. 38). The goal of education in this broader perspective is to provide beneficial changes in human character and personality, not just knowledge and practical tools, but to influence the kind of persons they become. Education is fundamentally a moral enterprise. It is, what John Dewey has called, the manifestation of humankind's responsibility to conserve, transmit, rectify, and expand "the heritage of values we have received" (Dewey 1934, p. 87). Education is the cultivation of humanity; fostering the capacity for critical examination of oneself and one's traditions and to see ourselves bound to all other human beings by ties of recognition and concern.

If this is the broad mission of education, how is this mission accomplished in global bioethics? It is argued elsewhere that bioethics should be a critical discourse that analyses and scrutinizes the current value systems pervasive in neoliberal globalization (Ten Have 2015). If many ethical issues arise because of these value systems, bioethics cannot simply reproduce this ideological context but should take a critical stance towards it and present alternatives. This is the double bind of global bioethics. It should critically review the context of globalization in which it has originated as well as the economic forces that are driving these processes of globalization. Because it has emerged in the context of globalization, the moral discourse seems already captured and determined with a preconceived value framework. Global bioethics, therefore, must emancipate from its sources and should adopt the Socratic task of being a gadfly or the Kantian role of philosophy as critical thinking rather than merely explaining and justifying current situations. Otherwise it will only serve to soften and humanize the neoliberal ideology that determines current globalization. This critical stance requires that global bioethics goes beyond the focus of traditional bioethics on individual autonomy and issues of science and technology and critically analyses the social, political, and economic context of healthcare and science. This critical refocusing is particularly important for bioethics education, now that in many countries education itself is significantly transformed into a commercial industry, remaking universities into businesses, students into customers, and academic research into an economic asset (Collini 2013). In the logic of marketization and quantification, the sole purpose of education is to provide graduates with capabilities that are demanded in the economy. If bioethics education accepts this logic, it will be anointing neoliberal ideology and will not develop an independent, critical stance. Global bioethics education needs to develop an alternative perspective based on ethical notions such as solidarity, human dignity, and social responsibility.

This book aims to address these challenges and to provide in-depth analyses of how they can be overcome. It will have three parts. The first is addressing the theoretical background of globalization and its implications for ethics education. The second part examines the goals of global ethics education, as well as the challenges that are presented in various cultural, social and political contexts, and economic inequalities. The third part presents and analyzes various examples, methods, and practices of global ethics education.

Moral Visions of Global Education

The first part of the book clarifies the global background and challenges of introducing and expanding bioethics education across the world. In the first chapter, Henk ten Have argues that the emerging discipline of global bioethics is inspired by the ideals of cosmopolitanism: the unity of humanity, solidarity, equality, openness to differences, and focus on what human beings have in common. These ideals consider each human being as a citizen of his or her own community or state (*polis*) as well as at the same time as a citizen of the world (*cosmos*). In the first, they are born; they share a common origin, language, and customs with co-citizens. In the second, they participate because they belong to humanity; all human beings share the same dignity and equality. Being a citizen of the world liberates the individual from captivity in categories such as culture, tradition, and community, but also gender and race. Humanism replaces communitarianism. Cosmopolitanism expresses the aspiration to live beyond specific, bounded horizons. It allows a broader solidarity without boundaries. The moral ideal is that human beings belong to a universal community ("humanity"); human well-being is not defined by a particular location, community, culture, or religion. Global citizens therefore have responsibilities toward other human beings, near or distant. Cosmopolitanism often uses the metaphor of expanding circles of moral concern taking into account more beings and entities as subjects of moral consideration. This chapter explores the implications of these ideals for bioethics education. Is it possible to develop global ethics education, contributing to the formation of global citizens concerned with global health and justice, assuming global responsibility to criticize structures of violence and inequity? This question is especially important since it is increasingly recognized that globalization is associated with rising injustices and inequalities.

In the subsequent chapter, Solomon Benatar further explores the new context for ethics and ethics education that is evident in a rapidly changing world and our threatened planet. The current focus on considerations of interpersonal ethics within an anthropocentric perspective on life should be extended to embrace considerations of global and ecological ethics within an ecocentric perspective on global and planetary health. The pathway to understanding and adapting to this new context includes promoting shifts in lifestyles from selfish hyperindividualism and wasteful consumerism toward cautious use of limited resources within an increasingly interdependent world in which the equal moral worth of all and sustainability are valued. Critical scholarly approaches to global politics and to the global political economy could facilitate such change and encourage iterative interactive processes instead of seeking conclusive definitive "scientific" solutions to all problems. Benatar hopes that this shift in perspective could be achieved firstly through sensitization to new and increasingly challenging ethical dilemmas, and then by encouraging rational thinking and action based on global and ecological considerations rather than on false economic dogma and the distorted workings of a market civilization. Moving ahead with these activities must begin with promotion of education, learning, and self-reflection to foster the widespread development of a global state of mind. Such a shift would require an expanded ethical discourse, with consideration of ethical dilemmas beyond human interpersonal relationships. These should include intrainstitutional and interinstitutional relationships, as well relational ethics between nations in a post-Westphalian world, and between humans and nature in an era now called the Anthropocene, to ensure survival on a planet undergoing entropy. The adverse effects of intense competition could be reduced by encouraging greater cooperation, linking security issues to global health and social justice issues, and using inspirational narratives and examples of moral imagination/moral leadership. These educational programs should be initiated in schools and become mandatory for all first-year college and university students. Such an agenda, argues Benatar, is feasible given human ingenuity, determination, resilience, and adequate resources, but will also require political and social will as its drivers.

Goals and Challenges of Global Ethics Education

In the second part of the book, practical issues regarding the introduction and implementation of teaching in global bioethics are addressed. Volnei Garrafa and Thiago Rocha da Cunha discuss goals of the global ethics education. The construction of the theory and practice of global ethics education can be undertaken with reference to various approaches. Only from the characterization of different views and interpretations of the theme, the aims (or goals) of this education may legitimately be defined. In this sense, this chapter will start with the necessary understanding of the term "global ethics" that gives epistemological support to it. Facing the dialectical tension between universalism and moral relativism, it will set global ethics in the context of recognition and appreciation of moral pluralism, taken as a cornerstone of the establishment of a global society truly united, free, and equal. Then it will explore the contradiction between the interpretation of teaching ethics as an individual phenomenon, directed exclusively to professional/private issues, or as structured academic proposal and directed to the wider public and collective interests, which consider not only the "me" and the "next," but also the distant and different, including future generations and forms of nonhuman life. A third topic that the authors address concerns the reasons for teaching global ethics from various points of view: academic, sustained in different theories and ethical propositions; and socio-political-economic, focusing on cultures, often disparate needs and processes and often antagonistic in the global context. The fourth and final part of the chapter has the task to present some of the different possibilities of classification and interpretation of the theme studied, among these: (a) macro and micro objectives, which can be changeable in terms of greater or lesser importance according to the issue or specific content to be worked on during historical moments that can also be varied; (b) direct or indirect goals to be achieved in line with the program and educational goals to which they are linked; (c) behavioral academic or social objectives, taking into account the sociocultural context in which the academic-educational process will be developed; (d) short, medium, and long-term time objectives, to contextualize the purposes to be achieved within a feasible timescale with different possibilities and limitations: human faculty resources, physical infrastructure where activities will be developed, reasons related to economic cost of the process, etc. In conclusion, this chapter presents an organic synthesis of the different parts developed, seeking to provide the reader with an integrated vision of the global ethics education goals.

The next chapter will consider the priorities that should prevail in the teaching of ethics in a globalized world. Renzo Pegoraro argues that globalization has created the favorable conditions for social, economic, and cultural integration in today's world. The development of information and communication technologies, that facilitate connecting people in different countries, has certainly made possible an overall interconnection among national and regional realities and made easier for people, goods, and services to move across borders. In this context, besides the many benefits of global range, several challenges—exacerbated by the diversity of perspectives in a society characterized by a pluralism of moral visions—are raised, requiring a specific reflection. These challenges involve general areas such as anthropology, medicine/healthcare, and ecology, as well as specific aspects: in a broad range, those linked to social justice fulfillment, guarantee safety, promote interreligious dialogue, and build peace; in a narrow range, those related to professional moral conduct, healthcare delivery, access of healthcare and equitable distribution of healthcare resources, and provision of treatment. Therefore, new priorities arise also in the teaching of ethics and bioethics in a globalized world. Pegoraro identified the following. First, "global conscience": we need to provide moral knowledge, skills, and attitudes, developing consciences to have a new ethical awareness, enabling first to recognize and then to face the challenge posed by the globalized world. The second is "universal values": we must spread universal values, particularly anthropological values such as human dignity, referring to those documents that have a certain consensus, such as the *Universal Declaration on Bioethics and Human Rights* (UNESCO 2005), aiming at these be recognized in every country (mainly in those countries where human life is threatened, and systematic violation of fundamental rights is permitted). The third priority is "common responsibility": rethink and propose again responsibility in seeking a sustainable and integral development. According to Pegoraro, common (or global) responsibility must take into account both promotion of human beings' dignity and preservation of creation, showing how respect for human life, justice, peace, and respect for the environment are absolutely interconnected themes (integral ecology).

In Chap. 5, Leonardo de Castro and Isidro Valero reexamine the four-principle approach to biomedical ethics in the context of ethics education in general and in relation to possible ethics discourse within a community of inquiry in particular. A community of inquiry is the setting for learning and education in philosophy for children. This community enables children to acquire critical thinking and other skills as part of democratic education. The use (or misuse) of the four principles approach tends to contribute to a practice that limits critical thinking skills because of the constraints on the conceptual tools that tend to be used. It has also had the effect of promoting conceptual ambiguity by encouraging the use of limited conceptual molds, thus giving rise to the possibility of multiple interpretations among diverse users. While recognizing the continuing appeal of the approach as a conceptual tool for ethical decision-making, this chapter brings out the limitations that need to be overcome in order to promote the clarity that the four principles approach is meant to possess. The argument of De Castro and Valero shows that the current methodology of mainstream bioethics is insufficient to address the challenges of contemporary globalization and the specific global problems that are generated by neoliberal ideology.

Ayesha Ahmad takes on the issues of interculturality and cultural competence in the following chapter. In light of contemporary humanitarian crises resulting from conflict, the demands on global bioethics as a resource for analysis and critique about cultural challenges are increasingly fraught, especially when situated in pluralistic religious and historical discourses. By virtue of its nature, humanitarianism creates a cultural encounter and this is evident on several levels. The dominant scientific medicine framework is exported across boundaries and implemented in very different contexts to that where it originated and formed a structure. Furthermore, the humanitarian physician is bracketed from local cultural understandings of the surrounding world. Developing a cultural lens that critiques these nuances is essential for practicing medicine in a global context especially during humanitarian crises. This chapter also highlights the need for cultural competence in relation to cultural conceptualizations of personhood. In order to analyze this neglected aspect of global bioethics, the chapter uses the case study of Post-Traumatic Stress Disorder during the global mental health movement and transcultural psychiatry debate. The chapter then begs a normative question about what does it mean to possess a differentiated cultural body and based on the findings of the chapter, it identifies crucial factors that need to be taken into account while establishing standards for cultural competence.

The last chapter in this part of the book examines the resources for global ethics education. Aimee Zellers provides helpful information about the various types of resources that are currently available for ethics education. Zellers argues that given the extent to which our lives are affected by what goes on around the world, it is time to rethink ethics education. As seen with SARS and Ebola, infectious diseases can be carried from one continent to the next. And as demonstrated by Chernobyl and Fukushima, environmental disasters in one area can have long-lasting impact on other peoples and places. In many respects global *is* local. Furthermore, we have a great deal to gain by expanding our horizons. We see this in bioethics, for example. Ilhan Ilkilic's work on the impact of culture and traditions on truth-telling calls into question our assumptions around honesty in doctor-patient relationships. Cecilia Wee shows in her work on Confucian Bioethics that bringing role morality into the discussion of acts of omission vs. commission could reshape the euthanasia debate. And Peter Omonzejele's analysis of vulnerable populations in Nigeria reveals some fundamental problems with informed consent in human subject research. Their contributions offer valuable perspectives on important bioethical issues. Assumptions shaped by the Western worldview have taken us a long way but there is a lot to be gained by taking a wider – more global – view. To help make that happen, this chapter surveys resources available in a global ethics education. Practitioners as well as educators will find resources ranging over theory and practice. This includes written and electronic resources as well as DVDs (documentaries and feature films) and videos. The quality of available resources makes it clear that the time has come for ethics to embrace a global perspective.

Practices of Global Ethics Education

The third and last part of this book includes four chapters that analyze practical tools that are often used in bioethics education, i.e., movies, stories, theater, and cases.

In Chap. 8, Jan Helge Solbakk presents his reflections on war veterans and refugees as survivors. An underlying assumption in this chapter is that as *survivors,* these conflict-ridden groups of people have to live through emotional upheavals of a similar kind and are faced with some of the same moral quandaries pertaining to the actual or perceived sense of homelessness. Recent reports from the United Nations indicate that there are around 43 million victims of war and conflicts who live displaced from their homes; of which more than 15 million have had to flee their own countries, while 27 million live as refugees in their own countries. The total number of war veterans worldwide has also reached enormous figures. The World Veterans Federation represents some 25–30 million veterans worldwide, and as of 2014, there were approximately 22.5 million war veterans only in the USA. Of about 33% of literally homeless males in the USA are veterans, and if one adds to that figure the number of war veterans suffering from a *sense* of homelessness—of living internally displaced lives—the reach of this problem becomes even more evident. The aim of this chapter is to illustrate how arts and humanities may be used as instruments to didactically reflect on and address these global problems. The arts sources that will be made use of are several of the tragedies of the ancient Greek playwrights Euripides (*Antigone*, *Ajax*, and *Philoctetes*) and Sophocles (*Heracles*,

Helen, *Orestes*, *Andromache*, *Hecuba*, and *The Trojan Women*). Solbakk is using the ancient poetic sources in combination with contemporary cinematic dramas of the problem of homecoming of war veterans and refugees.

Terry Maksymowych in the subsequent chapter examines how the arts can be incorporated into the bioethics curriculum. Traditional approaches to teaching bioethics courses include, among other things, case studies from clinical or research ethics, chapters on patient autonomy and beneficence, conflicts of interest, and the dangers of hidden prejudice. All of these are essential to a good bioethics course. However, incorporating the arts—whether it is a poem, short story, film, or a painting or play—enhances the curriculum and encourages students to see differently. The arts have the power to startle, to challenge beliefs, and to invite people to see the world through others' eyes. In our culture, the arts are often relegated to the province of entertainment, but they can be powerful teaching tools, particularly in a field of study in which empathy is highly valued. The beautiful and haunting poem, "Monet Refuses the Operation" helps students see the world from a disabled patient's perspective, where the quality of life may be interpreted quite differently from that of the physician's usual experience. Hawthorne's short story "The Birthmark" illustrates the dangers of scientific research without sensitivity to the needs and well-being of the patient, as well as the themes of physical enhancement and perfectionism. The HBO movie "Miss Evers' Boys" puts faces and personalities to those physicians, nurses, and patients involved in the infamous Tuskegee Study. Another Hollywood film, "GATTACA," explores the issues of genetic engineering and privacy. The paintings of Van Gogh and O'Keefe, the poetry of Coleridge and Byron, the novels of Woolf and the music of Schumann, offer glimpses into the minds of creative geniuses struggling with mental illness. Every culture has stories, art, music, and dance, and these can be used to speak to students in introductory or advanced bioethics courses. In science and humanities courses, they can be used to introduce bioethical dilemmas and promote discussion. An instructor can use examples from the students' own culture or expand their experience by introducing the arts and particular ethical issues of other cultures. Maksymowych in her chapter focuses on the methodology of incorporating the arts into bioethics classrooms, using examples from various cultural traditions.

In Chap. 10, Flavio Paranhos reflects on the cinema as an instrument for teaching and learning global ethics. He argues that the teaching of ethics has certain peculiarities that, while presenting themselves as difficulties, can also be an exciting challenge. The main concern of a course in bioethics is what exactly is expected as a result of it. Ethics being what ought to be done, those who study ethics will become experts on that which ought to be done. Certainly, things do not work quite that way. What a teacher of ethics can aspire to be the result of lectures is instigating the critical spirit of students in such a way that they will be able to never conform to what is given. Such disposition of spirit is even more important if we consider the ethics globally. Student and teacher must do the exercise of questioning things as they are not just in their backyard (or in a hospital ward), but on the entire planet. The position of those who are dedicated to the teaching and learning of ethics should never be comfortable. Following the existentialists, one needs to take positions, assume responsibility for what happens in the world. In this sense, the arts and humanities in general, and the cinema in particular are powerful instruments to sharpen the critical spirit of both students and the teacher; from the interaction between them, what follows a movie session in the classroom. Directors such as Woody Allen, Ingmar Bergman, Luis Bunuel, Akira Kurosawa, and Andrei Tarkovsky, among others, have works that are excellent tools for classroom discussions. To cite an example, *Crimes and Misdemeanors*, a 1989 film by Woody Allen, allows us to confront ourselves with the greatest weakness of the human being, which is the infinite ability to forgive himself. This is a start well suited to a course whose main objective is to take the students out of their comfort zone. If I cannot trust even myself, what should be my relation to everything around me I consider acceptable? The answer may not be at the end of the course, but certainly many other questions will join this one.

The last chapter of this book highlights the role of case studies in global ethics education. Willem Hoffman points out that effective global ethics educators should have a high-level awareness and application of educational principles and methods to create optimal contexts for learners/students to internalize ethical skills. Case studies are widely used educational instruments in this regard. This chapter addresses the following main areas: (1) the case study concept; (2) the educational aims of case studies; (3) the case study content; (4) the format of case study presentations; and (5) the scheduling of case studies. The first section of this chapter describes the definitions of the concept case study with reference to its understanding in various academic disciplines. The second section describes the educational aims of case studies. It facilitates the following educational outcomes: (i) self-reflection on personal values and value positions; (ii) identification and understanding of ethics principles; (iii) appreciation of ethics challenges; (iv) critical thinking skills; and (v) contextual critical reasoning. The third section focuses on case study content. The content must be congruent with the offering's overall learning aims. Also, it must be congruent with the general educational objectives to facilitate increasing levels of skills to (i) remember (to know facts, concepts, and procedures); (ii) understand (to explain and interpret); (iii) apply (to transfer knowledge and skills to real-life contexts); (iv) analyze (to differentiate and determine interactions and patterns); (v) evaluate (to make criteria-based judgments); and (vi) create (to plan or produce new patterns/structures). The content should ideally use authentic/realistic scenarios from a variety of local and international contexts, while its difficulty level should progress from simple to complex as the educational offering proceeds. Lastly, a variety of case study content sources can be utilized, namely casebooks, audiovisual material, news reports, formal literature/documents, and self-developed material. The fourth section focuses on case study presentation format. The main presentation modes are texts, verbal narratives, and audiovisual material, while an educator-facilitated approach or a learner/student-based self-guided approach can be used. Hoffmann, in the last section of his chapter, focuses on the scheduling of case studies. At the start of a session it is used as an icebreaker or moral game to raise ethical awareness, while during a session it illustrates concepts, stimulates reflection, and develops analysis skills. At the end of a session, it is useful for concept integration, to stimulate reflection, and for formative assessment. Case studies are used as home-based work during the course of formative assessment, personal reflection, and analysis skills development. Lastly, at end of the course, it is useful for summative assessment.

Henk ten Have

References

Collini, S. 2013. Sold Out. *London Review of Books*, October 24, 3–12.

Dewey, J. 1934. *A Common Faith*. New Haven and London: Yale University Press.

Mishra, P. 2017. *Age of Anger. A History of the Present*. Milton Keynes, UK: Allen Lane.

Rousseau, J.J. 1979. *Émile or on education*. Trans. Allan Bloom. New York: Basic Books (originally 1762).

Ten Have, H.A.M.J. ed. 2015. *Bioethics Education in a Global Perspective. Challenges in Global Bioethics*. Dordrecht: Springer.

Ten Have, H. 2016. *Global Bioethics. An Introduction*. London/New York: Routledge.

UNESCO. 2005. *Universal Declaration on Bioethics and Human Rights*. Paris: UNESCO.

Part I
Moral Visions of Global Education

Chapter 1
Cosmopolitanism and Educating the Citizen of the World

Henk ten Have

Abstract The emerging discipline of global bioethics is inspired by the ideals of cosmopolitanism: the unity of humanity, solidarity, equality, openness to differences, and focus on what human beings have in common. These ideals consider each human being as a citizen of his or her own community or state (*polis*) as well as at the same time as citizen of the world (*cosmos*). In the first, they are born; they share a common origin, language and customs with co-citizens. In the second, they participate because they belong to humanity; all human beings share the same dignity and equality. Being citizen of the world liberates the individual from captivity in categories such as culture, tradition, and community, but also gender and race. Humanism replaces communitarianism. Cosmopolitanism expresses the aspiration to live beyond specific, bounded horizons. It allows a broader solidarity without boundaries. The moral ideal is that human beings belong to a universal community ('humanity'); human well-being is not defined by a particular location, community, culture or religion. Global citizens therefore have responsibilities towards other human beings, near or distant. Cosmopolitanism often uses the metaphor of expanding circles of moral concern taking into account more beings and entities as subjects of moral consideration. This chapter will explore the implications of these ideals for bioethics education. Is it possible to develop global ethics education, contributing to the formation of global citizens concerned with global health and justice, assuming global responsibility to criticize structures of violence and inequity? This question is especially important since it is increasingly recognized that globalization is associated with rising injustices and inequalities.

H. ten Have (✉)
Center for Healthcare Ethics, Duquesne University in Pittsburgh, Pittsburgh, PA, USA
e-mail: tenhaveh@duq.edu

H. ten Have (ed.), *Global Education in Bioethics*, Advancing Global Bioethics 10,
https://doi.org/10.1007/978-3-319-78984-2_1

1.1 Introduction

The British Broadcasting Corporation undertakes every year a survey among citizens of eighteen countries to explore how they identify themselves. Do they regard themselves as citizens of the world in the first place or as national citizens? Remarkably, in 2016 more than 51% of the world population regarded itself as citizen of the world, up from 45% in 2001. In Nigeria the proportion is 73%, in China 71%, in Peru 70% and India 67%. In emerging economies, the majority of the population identifies itself as citizen of the world. In industrialized nations the trend is in the other direction. In the United Kingdom and the United States only a minority (40–45%) defines itself from a global perspective. In Germany only 30% see themselves as global citizens (down from 60% in 2002). The strongest sense of national citizenship exists in Russia; only 25% of all respondents regard themselves as citizens of the world (Grimley 2016).

This information is interesting. Of course, it is not clear what exactly people understand with the notion of 'citizen of the world.' It can primarily focus on economic issues; there is free trade across borders and people can assume they are participating in or benefiting from this trade system; especially people in developing countries can feel that they profit economically, that poverty is declining and the burden of disease diminishing. People can also relate the notion to the awareness that humanity is facing similar problems of climate change and inequality that can only be addressed by common action. Global citizenship for many is first of all experienced through social media and communication. One may feel a strong sense of global citizenship through being continuously connected with family, friends and colleagues, and even total strangers. Finally, global citizenship is experienced every day through mobility. It is easy to travel everywhere without significant obstacles from borders. Foreign countries are not really foreign anymore. Many people are migrating. The other side is that refugees are a growing issue but not necessarily a problem in an era of globalization.

However, the information indicates problems and changes. Western countries that have initiated the process of globalization, and have most benefitted from this process, are now engaging in activities of de-globalization. Populist movements in several countries blame the process of globalization for many problems and challenges in their countries. Instead of growing cooperation and solidarity there are an increasing number of self-interested autocratic regimes. Narcissistic leaders denigrate human rights discourse. This is not solely a Western phenomenon, but it is remarkable that the idea that we are all connected as inhabitants of our planet is now more shared by people in developing countries than in at least some of the developed world.

1.2 Cosmopolitanism

The notion of world citizenship is part of the philosophy of cosmopolitanism. Basically, this is the idea that there is a global moral community. All human beings are members of that community since they share common values and responsibilities (Ten Have 2016, p. 47–49; p. 107–109). Since Stoic philosophy it has been acknowledged that human beings are born, and therefore rooted, in a specific place; they are localized in a native community and state. They share a common origin, language and culture with fellow national citizens. At the same time they are inhabitants of the same planet; they are situated in a similar space. They share the same dignity and equality as members of humanity. As citizen of the world they can transcend their localization and boundedness of culture, tradition, community and history. Being born in a specific place and within a particular culture with its often restrictive and traditional customs, laws and morals can be overcome in cosmopolitan perspective. Cosmopolitanism is the aspiration to live beyond bounded horizons. Human beings are not defined, and should not be defined by a particular location, community, culture or religion. From this perspective, much older than the modern processes of globalization, boundaries have no moral significance; the focus should be on what human beings have in common. Cosmopolitanism expresses therefore the moral ideal of the unity of humanity. There is a universal community that includes the whole of humanity. All human beings therefore have equal moral status.

Cosmopolitanism is associated with the process of globalization, although it is contested. For critical scholars being a citizen of the world is an easy identification but it has no content and implication. It is merely a metaphor without substance since there is no world community to which these citizens belong. Nonetheless, it can be argued that during the last few decades globalization has had a transforming impact. First, the autonomy of nation states has significantly diminished, the body of international law has substantially grown, risks have multiplied and require global responses, and there has been an enormous growth of global organizations, especially non-governmental organizations. Thus there has been objective cosmopolitization. Second, at the political level, even anti-globalization politicians and actors have to acknowledge that the notion of sovereignty has changed. Individual countries cannot change much without cooperating at a global level. They may think they can act alone in their interests (for example ignore the issue of climate change and re-open coal mines) but climate issues come back to them, and will hurt their own citizens who are often not in agreement with current policies. Political activities nowadays are frequently on the agenda of civil society. Communities of citizens undertake global activities, putting pressure on traditional political authorities. Third, and most importantly, there is an emerging global consciousness. People know, feel, and experience that they live in a common world. Especially young people and students have grown up in an interconnected and global world. They live in what Milgram has called "a small world."(Milgram 1967) With new communication technologies and social media, people all over the world are closely connected (with six degrees of separation) while social networks bring people even closer together.

1.3 Global Consciousness

Global consciousness started to develop when humanity was confronted with the image of the globe. Pictures from the Apollo 11 spacecraft on its way to the first moon walk, taken shortly after lift off on 16 July 1969 from a distance of 180,000 km showed the Earth as a lonely planet in outer space. Although parts of the globe were covered by clouds the images were very clear with Africa in the center and parts of Europe, the Middle East and Western Asia clearly visible. Later images taken from the Apollo 17 in December 1972 were even more impressive. They showed the Earth in full as a perfectly rounded globe with recognizable continents. These images, called the "blue marbles," became the symbols of environmentalism. Human beings had never been able to see the planet that they inhabit. For the general public the confrontation with the image of the planet or 'spaceship' Earth made all aware how vulnerable and isolated the earth really is. The planet is now fully explored; all places in the world are known; there no longer is any 'terra incognita.' Our planet is a finite one. It is only a tiny part of an immense universe amidst an unknown number of other planets. If this is the home of humanity, all human beings share the same fate. They all should have the same concerns about the future. If this planet is ruined, every human being will suffer and perish. It is interesting that Van Rensselaer Potter, who introduced the notion of bioethics as a new and broader approach to medical ethics, used the Apollo 11 photo for the cover of his book *Bioethics. Bridge to the Future* (Potter 1971). Potter wanted to focus ethical attention on planetary and environmental issues rather than individual medical problems.

One of the effects of the images of Earth is that the 'global' is associated with space. The globe is out there, at a distance from us within an infinite atmosphere. It is also separated from us. It refers to a world apart from human life. It can be an object of contemplation, admiration or discontent but it determines the local place that is our habitat. Globalization has the same characteristics. Its processes influence and transform the local place. While the global is homogeneous and universal, the local produces heterogeneity. Global forces are converted into specific practices and activities through the agency of local forces, either building alternatives or resisting global forces. This view of the interaction between global and local often assumes that the local is subjected to the global, or that the global is more or less imposed on the local. Since the global is associated with space, borders are irrelevant and globalization is similar to deterritorialization, local places are sites of application and implementation rather than locations where the global is in fact constituted, produced and invented (Massey 2004).

1.4 Globalization: The Reality

According to liberation theologian Leonardo Boff the reality of life in 1978 for the majority of people in Latin America was the following: "They are living and dying amid inhuman living conditions: malnutrition; a high infant mortality rate; endemic diseases; low income; unemployment; lack of social security; lack of health care, hospitals, schools and housing facilities. In short, they lack all the basic necessities that might ensure some minimum of human dignity" (Boff 1978, p. 268). We can wonder how different the situation is in 2017. The continent now has more middle-class people than poor. But 130 million people have not benefitted from globalization. Despite economic growth one in five persons lives in chronic poverty (Rigolini and Vakis 2017). One third of Latin Americans survive on $1 to $2 per day.

More than 52 million people (10% of the population) suffer from malnutrition or hunger. Approximately 30% of the population has no access to health care for economic reasons. One out of two inhabitants is excluded form social security systems (ILO 1999). These data indicate that globalization has indeed improved the fate of many. But the conditions mentioned by Boff continue to exist for millions of people in the continent. Similar findings occur in other continents. In Africa the average life expectancy is 53 years. Every minute a child dies of malaria, a preventable and treatable disease. Approximately 20% of all hungry people in the world live in Africa. The conclusion of these findings often is that globalization has positive effects. It improves the life of many people in many countries. This is especially true for global health: it has led to enormous improvement of health and life expectancy. In developed countries the mortality rates for infectious disease and cardiovascular disease have significantly decreased. In developing countries, especially child mortality is much lower. The global average age of death increased from 46.7 in 1990 to 59.3 in 2013. The global average life expectancy for both sexes increased from 65.3 years in 1990 to 71.5 years in 2013 (Global Burden of Disease Study 2015). In general, globalization has been associated with international expansion of medical research, wider availability of medication and treatment, and growing international cooperation, aid and exchange. However, this progress has been unequal. It has produced paradoxes of global health.

The first paradox is that there is growing disparity in health. Life expectancy at birth differs substantially among countries. In Sierra Leone it is 50.1 years, in Nigeria 54.5 but in the Netherlands people can expect to live 81.9. In Japan the life expectancy is longest: 83.7 years. Also, within countries, life expectancy can differ. Michael Marmot (2015) has demonstrated that in a city such as Glasgow there is a 28 year gap in life expectancy between living in a poor part of the city compared to a richer neighborhood. Similar disparities exists in many other dimensions of healthcare, for example, access to essential medication or prevalence of diseases. A major concern is the unequal distribution of healthcare workers. The majority of them are working in the Americas and Europe, areas where the burden of disease is the lowest on a global level (WHO 2006).

The second paradox is that improvement of global health leads to progressive deterioration of the natural world. In order to feed a growing human population (and thus sustain human health), intensive agriculture has enormously expanded. This expansion requires more land to grow food, and thus in many developing countries increasing deforestation. Intensification of agriculture demands more use of pesticides and herbicides. It is also associated with monocultures, making crops more vulnerable to diseases and pests. Deforestation furthermore is related to the emergence of new viral diseases, producing an increasing number of pandemics.

The third paradox is the increase of global ethical problems such as human trafficking, organ trade, brain and care drain, and bio-piracy. Such problems are new challenges for bioethics. They are different from the earlier generation of ethical problems since they are not related to developments in science and technology or medical treatment. They are also not individual problems that can be addressed by empowering patients or limiting medical interventions. As global problems they require cooperation and collective action. First of all, these problems are the result of particular policies and practices of globalization. Bioethics should therefore focus on these global processes and their underlying value structure.

1.5 The Problem of Inequality

How can we explain the paradoxes of global health? Globalization has produced in a relatively short time an enormous improvement of global health. This is an impressive benefit. However, the point is: not for everybody, and not everywhere. Some are reaping the benefits of the processes of globalization while others do not; they are marginalized, neglected, and harmed by the same processes. The fundamental moral problem that underlies the paradoxes of global health is inequality.

Inequality is not a new problem. In human history there has always been inequality in income and wealth, gender, race and ethnicity, class, power and social status. But globalization has significantly exacerbated economic inequality within and between countries. Types of inequality are also related. Economic inequality is associated with other types of inequality, particularly health inequality and inequality in environmental degradation. Inequality is often a symptom of vulnerability. The fact that the world has become increasingly interconnected and interdependent has created a sense of mutual vulnerability. In the words of the Director-General of the World Health Organization: "Vulnerability is universal." (WHO 2007, p. 2). Being vulnerable is often the result of a range of social, economic and political conditions, and therefore beyond the power and control of individuals. Processes of globalization have resulted in a world that is not only creating more and new threats, but they have also undermined the traditional protection mechanisms (social security and welfare systems, family support systems) so that the abilities of individuals and communities to cope with threats are eroded. Entire categories of people are disenfranchised, powerless and voiceless (UN 2003). It is clear that this interpretation of vulnerability as global phenomenon is at odds with mainstream bioethics' framing of vulnerability as an individual affair.

Recent data of Oxfam Novib informs us that the 85 wealthiest persons in the world are as rich as the 3.5 billion poorest people (Oxfam 2014). In the U.S. the top 1% of the population earned 23% of all income in the country (this was 11% in 1944 but 24% in 1928) (Stiglitz 2012). These differences in wealth and income have been growing rapidly. In 1979 the average income of the top 0.1% was 50 times greater than the average income of the bottom 90%. In 2010 the ratio was 164 times that of the average of the bottom 90%. In the U.S. the wealthiest 1% controls 40% of the wealth of the nation; they also receive one-fifth of the total income. For almost three decades the incomes of the majority of the American population have been stable; productivity increased but only the CEOs benefitted; their annual compensation in

2010 was 243 times higher than that of the average worker. Minority populations were heavily damaged. As Joseph Stiglitz pointed out: "Between 2005 and 2009, the typical African American household has lost 53 percent of its wealth…, the average Hispanic household has lost 66 percent of its wealth." (Stiglitz 2012, p. 13).

Nowadays, inequality is regarded as a major social and political problem (Greig et al. 2007). It can also be argued that it is the basic contemporary problem for ethics, and especially bioethics. Many have ignored and de-prioritized it for a long time as the inevitable consequence of globalization, arguing that it will diminish in the long run as long as priority is given to the market ideology. Today it is no longer possible to deny that it exists and is growing. The market is not solving this problem but amplifying it. Recent polls in the United States show that two-thirds of the population is dissatisfied with the distribution of wealth and income (Riffkin 2014). In the United Kingdom this percentage is even higher: 82%. Many people feel that something is fundamentally wrong and unjust; the economic system has brought them into an undignified situation. Inequality has many negative effects not only at the global level but also within countries and regions. It leads to political and social instability, undermines democracy and social cohesion, encourages lack of solidarity, and erodes a sense of community, as well as fair play and justice. Stiglitz summarizes the injustice in one sentence: "We created for the banks … a much stronger safety net than we created for poor Americans" (Stiglitz 2012, p. 74).

Neoliberal ideology has for a long time argued that the free market is a self-regulating and irresistible force. The world has become one single global market. The market is the most fair and efficient form of social organization. It allows individual liberty and freedom to flourish. Competition is the core value. There is no need for regulation or protection. In the end everybody will benefit since these benefits will 'trickle down.' Inequalities will only be transitional ('a rising tide will lift all boats') (ten Have 2016, p. 67 ff). For most people it is obvious now that this ideological discourse is wrong. Neoliberal policies have only multiplied vulnerabilities and inequalities. For example, it has restricted access to medication particularly in developing countries. Only the collective action of social movements has forced governments and international trade organizations to acknowledge that public health should be more important then trade, and that the right to health should be implemented; it is immoral to let people die because effective drugs are too expensive (ten Have 2016, p. 126). Even economists themselves are now deconstructing the myths of neoliberal ideology, arguing that wealth is not the result of hard labor but is inherited and the result of returns on capital; that the trickle down effect does not exist at all; that the market is not efficient but depends on subsidies and creates monopolies so that competition in practice is limited (Piketti 2014; Parkin et al. 2013).

The most important conclusion of critical studies is that change is possible. The free market is not free at all and market forces are not autonomous. The market has been shaped by governmental policies. It is not beyond our control. Politics has promoted neoliberal practices and policies, and has therefore also created the inequalities. History tells us that inequality is not a fact of life. Between 1945 and 1970 there was a marked reduction of inequality in Europe because of deliberate policies of expansion of the welfare state and social provisions, financed by progressive income taxation. Anthony Atkinson argues that changing the tax structure (lower income tax) sharply contributed to the rise of inequality since the 1980s. After decades of declining, inequality has risen in many countries that started to change the social and economic structures, privatizing public goods such as healthcare and education, and dismantling social safety nets. This 'inequality turn' in the 1980s was not a natural event but the result of deliberate neoliberal policies (Atkinson 2015).

1.6 Inequality and Education

Globalization has provided a new context for education (Burbules and Torres 2000; Stromquist and Monkman 2000; Noddings, 2005; Hansen 2011). On the one hand it has generated a positive response with a focus on global citizenship and cosmopolitan initiatives. It has expanded the consciousness of students and scholars as subjects connected to others on the planet and as members of the global community. On the other hand, the idea and practice of education has been thoroughly transformed by the neoliberal ideology. Education became a tradable commodity on the global market with its inclusion in the General Agreement on Trades and Services (GATT). Rather than being a public good, it was transformed into a product that can be marketed by private corporations in competition with public educational institutions. Universities have to be run like for-profit companies. Academic scholars are entrepreneurs, students are consumers, education is training for jobs, and university presidents act like CEOs (Giroux 2014). Globalizing efforts in science and knowledge are often driven by the neoliberal model of globalization. International outreach is justified with the argument that it will claim revenues from a growing market and that students need to be prepared for this global market, rather than with references to global citizenship, inequality or injustices (Haigh 2008).

A neoliberal interpretation of education has at least two impacts. First, it emphasizes economic skills and competencies. A major concern of educational institutions is that students will become efficient and competent workers. In the present context of a globalized world this means that students have to acquire certain competencies. They have to know various cultures and how to operate within them. The Swiss Consulting Group (2002, p. 4) for example defines global competence as: "the capacity of an individual or team to parachute into any country and get the job done while respecting cultural pathways." This competence is purely instrumental. It does not necessarily have any interest in globalization. It also does not imply the idea of global citizenship. It just reflects a business model. Investing in professional development and being educated in globalization will increase productivity. University faculty however is critical of this instrumental view of internationalization: not economic self-interest but ethical orientation will be required. Engaging and participating in global education demand concerns with inequality (Hanson and McNeil 2012). Second, the language of competence favors an individualistic approach. Competence is something an individual can acquire, possess, and demonstrate. Competence is a quality to be achieved; individual learners are initially empty vessels. The emphasis therefore is on individualist approaches to learning. It is assumed that learning processes are autonomous; there are individual actors; knowledge is a commodity and private property. In other words, learning is an individual enterprise rather than a collective one. It is not a social activity based on participation and interaction, and linked to a specific context (Eichbaum 2015). In this approach competence is context-free; it is not bounded to time and space; global competencies can be applied everywhere. These views of course are very neoliberal. The emphasis on individualism is not only itself determined by the specific context of Western liberalism, but also typical for free market ideology. It assumes the priority of consumers over citizens. Human beings are primarily regarded as autonomous choosers that want to maximize personal preferences. There is no moral commitment to global humanity (Dill 2012; Lingard 2012).

The neoliberal approach to education is also affecting medical education. Kenneth Ludmerer in his recent study of graduate medical education concludes that "…the education of tomorrow's doctors is endangered by our country's commercialized system of health…" (Ludmerer 2015, p. xii). He examines the history of the residency system; the clinical years during which medical graduates specialize in a particular area. This is the pride of American healthcare established in the last century by William Osler. In this period not only knowledge and skills are learned but also the values, attitudes and behaviors of young physicians are shaped. First of all, the emphasis is on learning by doing. Its inspiration is Dewey's philosophy of education. According to the traditional values guiding this period of medical eduction, critical intellectual qualities, habits of thought should be acquired. Education rather than training should be pursued. This implies the "ability to adapt to the future" (Ludmerer 2015, p. 174). It is not just knowing what to do but understanding why it should be done. A second important idea in this educational system is learning as a communal activity. Students are engaged in grand rounds, morning reports, and interactive discussions with teachers and among themselves. Thirdly, it also emphasizes the moral dimension of patient care. The patient comes first and care is a duty. However, Ludmerer shows how these qualities of the educational environment are declining nowadays. The development of hospitals into large medical centers is delivering increasing patient services. It promotes shorter stays in hospital, more patients admitted and quick discharge. These commercial drives are associated with an increase of burnout among residents, impersonal contexts, less supervision, focus on productivity and throughput. It puts more and more service duties on residents as cheap labor for hospitals. Patient care is regarded as "market commodity" (Ludmerer 2015, p. 231). There is little time for reflection and deliberation, or for relationships with inspiring teachers. The educational community is weakened. The basic problem of residency education is, according to Ludmerer, that learning and practicing medicine now take place "…in a commercialized health care environment that gave mainly lip service to "patient-centered care." (Ludmerer 2015, p. 325).

The emphasis on competencies in current medical education is also criticized. It has the advantage that competencies are measurable, so that it can be assessed whether medical students make progress. One can follow the gradual attainment of competences by evaluating the tasks that students perform. But being a medical doctor is more than implementing a set of 'doings' and performances. It is ultimately a matter of professional identity. In the end medical education should develop professional identity (Jarvis-Selinger et al. 2012).

1.7 Global Bioethics

The ideals of cosmopolitanism have inspired processes of globalization. At the same time, the way globalization has developed under the influence of neoliberal ideology has created problems such increasing inequality and vulnerability. It has also influenced and transformed ideas and practices of education. The underlying values of neoliberalism, however, are rarely addressed in mainstream bioethics discourse (Ten Have 2016). In order to critique these values, a new, broader and more encompassing ethical discourse will be necessary. Van Rensselaer Potter has argued: "The time has come to recognize that we can no longer examine medical options without considering ecological science and the larger problems of society on a global scale…" (Potter 1988, p. 2 and 78). The same is true for bioethics education: it will need a global perspective.

The global bioethics advocated by Van Rensselaer Potter is finally coming into existence (Potter 1988). As stated in a recent publication: it is critical for bioethics "to incorporate the realities of a globalised world, one with increasing disparities and power differences" (Ganguli Mitra and Biller-Andorno, 2013, p. 101). Mainstream bioethics is challenged to develop and expand its theoretical framework beyond the principles and approaches established in the 1970s. It also urges bioethics beyond its initial frame of reference that is heavily influenced by North American culture and ideology. A lot of theoretical work is currently done to develop such broader theoretical frameworks based on human rights, social justice, capabilities and global care ethics.

When bioethics discourse was initiated and expanded during the 1970s and 1980s the major moral challenges were related to the power of science and technology. How can patients be protected against medical interference and paternalism? How can citizens have more control over healthcare decisions? In what ways can patients' rights be defined and implemented? These questions have shaped the agenda and methodology of mainstream bioethics, especially in more developed countries. But in a global perspective, many citizens do not have access to modern science and technology. They are marginalized in a system that is increasingly privatized and commercialized. They are exploited in clinical research projects since it is their only chance to receive treatment and care. It is obvious that in this perspective, especially since the 1990s the major moral challenges have changed. It is no longer the power of science and technology that produces ethical problems but the power of money. Healthcare, research, education, and even culture and religion are regarded as businesses that are competing for consumers.

When the major bioethical problems of today are produced by the dominance of neoliberal market ideology, bioethics should redefine itself as critical global discourse. Focusing attention on the social context will not be enough. Bioethics must argue for a reversal of priorities in policy and society: economic and financial considerations should serve the principles of human dignity and social justice, and no longer be ends in themselves. This implies specific strategies for social inclusion but also institutional support. It will be necessary to demonstrate more vigorous advocacy and activism, supplementing academic enquiry. Social inequalities and conditions that produce vulnerability are not beyond social and political control. It will also require that the voices of the disadvantaged, the deprived and the vulnerable are more often heard within the bioethical discourse, involving vulnerable groups in policy development and implementation. Global inequality and vulnerability are furthermore transforming the significance of cooperation. Forging global alliances and new networks of solidarity is the only way to address global threats. An individualistic perspective makes it impossible to address the root causes of global moral problems. Influencing and changing social conditions require what Fiona Robinson has called "collective capacity to act" (Robinson, 2011, p. 60).

1.8 Teaching Ethics in Times of Inequalities in a Global Society

The notion and practices of education have been transformed by free-market fundamentalism. Human beings are addressed as consuming and producing machines ready for the world as an enormous supermarket. This neoliberal approach to education is affecting our capacity to imagine another world. Henry Giroux argues that it specifically prevents us from imagining "a different and more critical mode of subjectivity and alternative mode of politics" (Giroux 2014, p. 14). The role of education no longer is "creating an informed citizenry," let alone cosmopolitanism (Giroux 2014, p. 21). There is no need for critical thought, knowledge, dialogue, and dissent.

Precisely for these reasons, ethics education in times of neoliberalism and inequality has a specific agenda. It should counter the neoliberal discourse and emphasize different values. It should introduce moral imagination, articulating the language of the social, solidarity, collective agency, public life, and democratic participation. Education should be promoted as UNESCO (2015) argues in a recent report, as a global common good and as a right. This implies that education itself has an ethical mission; it promotes learning for expanding justice, social equity, global solidarity, and respect for human dignity. It should be recognized as a public effort that is vital for democratic and civic life. This is why the humanities are needed. Studying and analyzing the human condition is necessary to promote the common good (Drakeman 2016).

What is the role of ethics in education against the backdrop of globalization? The discussion on global competencies is diverse but generally inspired by cosmopolitanism aimed at creating citizen of the world. If the purpose of education traditionally has been to prepare students for active citizenship in their own nation and culture, what will be its purpose in times of globalization? Global consciousness is not merely a matter of knowledge and skills but the result of an ethical transformation. Promoting global citizenship means promoting ways of thinking and acting that may transform current globalization. It is more than knowledge transfer. This kind of citizenship is associated with empathy for others, valuing diversity, social justice, environmental sustainability, and responsibility to work together for the betterment of the world (Quisumbing 2005; Golmohamad 2009; Reysen and Katzarska-Miller 2013).

Educating citizens of the world requires as a first step understanding cultural differences. Students should become aware of world history, of differences in geography and culture. They need to be open and aware of heterogeneous perspectives. That assumes that they understand their own culture and its dominant normative framework. In many countries citizens are more likely to believe that their culture is superior to that of others (Reimers 2009; Hunter et al. 2006). Awareness and critical thought about one's own culture and tradition is connected with a second step: situating oneself within a shared global community. This is recognition that one is a citizen of a nation as well as the world at the same time. It means recognizing that there are universal principles for every human being on earth as well as principles that are specific for particular countries, cultures and religions. Kwame Appiah has abstracted this challenge: cosmopolitanism is universality plus difference (Appiah 2008).

For Martha Nussbaum the goal of education is (and has always been) the "cultivation of humanity." Three capacities are essential for this goal. The first is the capacity for critical examination of oneself and one's traditions. Second, the ability to see oneself as human being bound to all other human beings by ties of recognition and concern. And third, narrative imagination: "the ability to think what it might be like to be in he shoes of a person different from oneself, to be an intelligent reader of that person's story, and to understand the emotions and wishes and desires that someone so placed might have" (Nussbaum 1997, p. 10–11; Nussbaum 2002). Cosmopolitanism in education therefore is more than cross-cultural competence; it is also the commitment to global values such as justice, solidarity, and social responsibility (Kennedy 2015). Furthermore, global citizenship, according to other authors, should also have practical implications. It should go beyond academic and theoretical commitment. It starts with awareness of our role in world (the global consciousness mentioned earlier), but this should lead to responsibility for the global common good, recognizing moral obligations to other people elsewhere. Global interdependence and solidarity finally should encourage participation in activities to ameliorate global problems (Banks 2004; Schattle 2008). Because cosmopolitanism implies a commitment to human rights and democracy, citizens of the world should undertake actions to counter the effects of neoliberal policies (Burbules and Torres 2000).

1.9 Implications for Bioethics Teaching

Globalisation and particularly increasing health disparities have encouraged more and more calls to rethink medical education. Usually medical education is not focused on global health and the fact that borders have disappeared. Doctors nowadays face a new set of problems also in the area of ethics. The educational system is not well prepared to tackle these global challenges (Frenk et al. 2010; Horton 2010; Wintrup 2015; Wass 2015; Wass and Southgate 2017).

Bioethics education, particularly at the global level is also confronted with several challenges. Bioethics teaching programs have mushroomed since the 1970s and 1980s. But it is not clear that this situation will not deteriorate under economic and political pressures on universities replacing experienced staff with temporary adjuncts and online courses. Ethics teaching is also regarded by policy-makers as a curious type of palliative remedy. Every time professionals infringe on important ethical norms, the need for ethics teaching is re-emphasized as the antidote. In response to a repeated cycle of cases of scientific misconduct and ethical problems concerning financial conflicts of interest, the National Institutes of Health and the National Science Foundation in the U.S. have required as of January 2010 that researchers funded by their grants must have received ethics education focused on promoting research integrity. Education in ethics is seen as a remedy against deficiencies in professional behavior. But it is obvious that the impact of bioethics education is limited if the systemic and structural causes of such misconduct are not addressed. Also, hardly any provisions and regulations are provided for ethics education so that a one-day online course can be sufficient to meet the requirements. Ironically, there is almost general agreement that bioethics education is very important for healthcare professionals; at the same time, it is not translated into efficient practical arrangements. Although bioethics teaching is practiced, in most countries it is not very impressive in terms of volume, time, and commitment. Persad et al. (2008) point out that in the U.S. bioethics education, although required, comprises only 1% of the medical school curriculum. Many educational activities are sporadic and occasional. In Europe most hospitals have only short-term educational initiatives instead of longer courses and programs, while nobody seems to take responsibility for the activities (Pegoraro and Putoto 2007). Moreover, there is a serious lack of qualified teachers. Not even half of the bioethics instructors in the U.S. have published a single article in bioethics (Persad et al. 2008). For many teachers of bioethics this is not their primary academic focus. The first challenge therefore is that the professed importance of bioethics education should not blind us for the frail and anemic status of programs in many settings.

Another challenge is related to bioethics education itself. It is exemplified in the enormous heterogeneity of the field. It is not at all clear what such education should entail. Within the same country, different types of programs are offered, didactic approaches and methods differ, the number of teaching hours has a wide range, and ethics courses are not scheduled in the same phases of the curriculum. Major controversies exist concerning the objectives, methods, content, and evaluation of teaching activities (Ten Have 2014; Carrese et al. 2015). However, this diversity does not imply that there is no consensus. Over the last few decades scholars have come to agree that certain approaches of teaching are preferable, for example, that there is a need for longitudinal and integrated programs, making ethics not an isolated, one-time event but part of daily care routine; there is a need for team teaching with close cooperation between ethicists and clinicians. Also a student-centered approach in bioethics education focused on active learning is preferable since it encourages critical thinking and reflection. Furthermore, there is agreement on the need for comparative studies. Developing teaching programs is often not informed by experiences elsewhere. In many cases the wheel is re-invented since there are few descriptive and analytic studies of specific programs published. Finally, more efforts are undertaken to define a common core for bioethics education, for example the core proposal in the United Kingdom, and the core curriculum launched by UNESCO (Stirrat et al. 2010; Ten Have 2008).

As argued earlier, this heterogeneity reflects two diverging views of bioethics education (Ten Have 2015). The question of what is good education requires a prior answer to the more fundamental question: why do we educate at all? Responding to this query one can observe two different philosophies of bioethics education. One pragmatic view regards ethics teaching as a way of learning skills for analyzing and resolving the ethical dilemmas that will confront health professionals in their future practices. The role of bioethics education therefore is limited. It should focus on what is practical and measurable. In this modest educational philosophy it is not realistic to expect that ethics education can create morally better physicians and scientists. After all, how can a limited number of courses bring about a change in behavior or character of health professionals? The primary objective therefore is to teach skills so that it will ultimately lead to better professional decisions. The other view is broader and bolder. In this philosophy, bioethics education is not merely focused on skills to improve decision-making but is basically a long-term effort to create better health professionals and scientists. It is aimed at character formation, integrity, and professional virtues. Rather than enhancing professional skills it aims to improve the professional him- or herself. Only in this way can bioethics teaching contribute to enhancing the quality of patient care. This broader philosophy is motivated by the fact that bioethics education was introduced and promoted to counteract dehumanizing and objectifying tendencies in contemporary medicine and health care. It is not just there to facilitate medical decision-making, but it should contribute to making medicine more humane. For this reason, bioethics education has a broader focus on the humanities, liberal arts, social sciences, and philosophy, so that medical activity is located within a wider human context.

It seems that the philosophy of bioethics education is increasingly moving towards this broader conception. While the focus on identifying and analyzing ethical issues has been characteristic for the early stages of bioethics education, at present there is more emphasis on how to influence students' attitudes, behaviors, and characters, emphasizing that the ultimate goal of bioethics education is to produce good health professionals and scientists (Goldie 2000). Good medical practice requires more than knowledge and skills. We expect health professionals to demonstrate good conduct and action. This is what education should train and nourish. The focus of bioethics education should therefore move beyond problem-solving and applying principles.

The need for a broader focus of bioethics education is even more necessary given the emergence of global bioethics as a consequence of processes of globalization. Nowadays, globalization is a major source of bioethical problems. Neoliberal market ideology is seriously damaging health and healthcare at the global level, creating many of the global problems mentioned before. Thus, the same source that has produced global bioethics is also generating the relevant global problems of today (Ten Have and Gordijn 2014). The impact of neoliberal globalisation is visible in all sectors of healthcare, medicine and research.

This context of globalisation clarifies the main mission of bioethics education nowadays and in the near future: bioethics should be a critical discourse that analyses and scrutinizes the current value systems pervasive in neoliberal globalization. If many ethical issues arise because of these value systems, bioethics cannot simply reproduce this ideological context but should take a critical stance towards it and present alternatives. This is the double bind of global bioethics. It has to critically review the context of globalization in which it has originated as well as the economical forces that are driving these processes of globalization. Because it has emerged in the context of globalization, the moral discourse seems already captured and determined within a preconceived value framework. Global bioethics therefore has to emancipate from its sources and should adopt the Socratic task of being a gadfly or the Kantian role of philosophy as critical thinking rather than merely explaining and justifying current situations. Otherwise it will only serve to soften and humanize the neoliberal ideology that determines current globalization. This critical stance requires that global bioethics goes beyond the focus of traditional bioethics on individual autonomy and issues of science and technology and critically analyses the social, political, and economic context of healthcare and science. This critical refocusing is particularly important for bioethics education, now that in many countries education itself is significantly transformed into a commercial industry, remaking universities into businesses, students into customers, and academic research into an economic asset (Collini 2013). In the logic of marketization and quantification, the sole purpose of education is to provide graduates with capabilities that are demanded in the economy. If bioethics education accepts this logic it will be anointing neoliberal ideology and will not develop an independent, critical stance.

Criticizing neoliberal market ideology requires a broader framework than the usual emphasis on individual autonomy. This current emphasis is convenient for the neoliberal perspective since it regards human beings first of all as individual rational decision-makers and consumers. What they need is information so that they can choose what they value or desire. But a human being alone is "an impossibility" (Taylor 1985, p. 8). Even economists nowadays argue that this conception of individual autonomy is an "anthropological monster" (Cohen 2012, p. 34). Regarding human beings as self-interested, self-determining subjects disregards the basic importance of cooperation, the interconnectedness of human beings, and the interrelations between human beings and the environment. Global bioethics education needs to develop an alternative perspective based on ethical notions such as solidarity, human dignity, and social responsibility. At the same time, the nature of global bioethical problems requires a re-orientation towards global values.

Bioethics education is a powerful means to counter the discourse of neoliberalism and transform this into a bioethical discourse focused on different global values. The ultimate purpose of education is 'metanoia,' a transformation of our views of ourselves, others, society and the planet. As citizens of the world this means going beyond micro-ethical issues related to decisions and actions of individual persons but rather focusing on the macro-ethical context. It also focuses on the social role of physicians and scientists who have moral responsibility towards society and the planet and take into consideration the common good of humankind. The emphasis of such education will be on transformation, a change in perspective, rather than knowledge and skills. Using global values such as human dignity, respect for diversity, social justice, social responsibility, vulnerability and protection of future generations, bioethics education can foster global imagination so that global ethical problems can be seen in a different light.

1.10 Conclusion

Markets are eroding moral values (Falk and Szech 2013). The neoliberal ideology promoting free market thinking and behavior has significantly transformed contemporary societies, and specifically healthcare and education. Mainstream bioethics since the 1970s has developed in association with neoliberal policies and practices. Bioethical discourse with its emphasis on personal autonomy and individual responsibility has encouraged consumerism and individualism. It has deflected critical analysis away from social, political and economic structures and contexts. Since many bioethical problems of today are global in scope and range, and are rooted in neoliberal approaches to health and education, this chapter has argued that the ethical discourse should change. It should focus on the neoliberal context and the values implicit in this context, offering alternative and different global values. These values such as solidarity, vulnerability and justice should be the core matter of bioethical teaching programs.

References

Appiah, K.A. 2008. Education for global citizenship. *Yearbook of the National Society for the Study of Education* 107 (1): 83–99.

Atkinson, A.B. 2015. *Inequality: What Can be Done?* Cambridge, MA/London: Harvard University Press.

Banks, J.A., ed. 2004. *Diversity and Citizen Education. Global Perspectives*. San Francisco: Jossey-Bass.

Boff, L. 1978. *Jesus Christ Liberator*. Maryknoll: Orbis Books.

Burbules, N.C., and C.A. Torres, eds. 2000. *Globalization and Education. Critical Perspectives*. New York/London: Routledge.

Carrese, J.A., J. Malek, K. Watson, et al. 2015. The essential role of medical ethics education in achieving professionalism: The Romanell report. *Academic Medicine* 90: 744–752.

Cohen, D. 2012. *Homo economicus, prophète (égaré) des temps nouveaux*. Paris: Albin Michel.

Collini, S. 2013. Sold out. *London Review of Books*. 35 (20): 3–12.

Dill, J.S. 2012. The moral education of global citizens. *Society* 49: 541–546.

Drakeman, D. 2016. Why we need the humanities. In *Life Science, Law and the Common Good.* New York: Palgrave Macmillan.
Eichbaum, Q. 2015. The problem with competencies in global health education. *Academic Medicine* 90 (4): 1–4.
Falk, A., and N. Szech. 2013. Morals and markets. *Science* 340: 707–711.
Frenk, J., L. Chen, Z.A. Bhutta, et al. 2010. Health professionals for a new century: Transforming education to strengthen health systems in an interdependent world. *The Lancet* 376: 1923–1958.
Ganguli Mitra, A., and N. Biller-Andorno. 2013. Vulnerability and exploitation in a globalized world. *International Journal of Feminist Approaches to Bioethics* 6 (1): 91–102.
Giroux, H.A. 2014. *Neoliberalism's War on Higher Education.* Chicago: Haymarket Books.
Global Burden of Disease Study 2013. 2015. Global, regional, and national incidence, prevalence, and years lived with disability for 301 acute and chronic diseases and injuries in 188 countries, 1990–2013: A systematic analysis for the Global Burden of Disease Study 2013. *The Lancet* 386 (9995): 743–800.
Goldie, J. 2000. Review of ethics curricula in undergraduate medical education. *Medical Education* 34 (2): 108–119.
Golmohamad, M. 2009. Education for world citizenship: Beyond national allegiance. *Educational Philosophy and Theory* 41 (4): 466–486.
Greig, A., D. Hulme, and M. Turner. 2007. Challenging global inequality. In *Development Theory and Practice in the 21st Century.* Houndmills: Palgrave Macmillan.
Grimley, N. 2016. Identity 2016: 'Global citizenship' rising, poll suggests. *BBC News*. 28 April 2016. http://www.bbc.com/news/world-36139904.
Haigh, M. 2008. Internationalisation, Planetary Citizenship and Higher Education Inc. *Compare: A Journal of Comparative Education* 38: 427–440.
Hansen, D.T. 2011. *The Teacher and the World. A Study of Cosmopolitanism as Education.* London/New York: Routledge.
Hanson, C., and B. McNeil. 2012. Faculty understanding and implementation of internationalization and global citizenship. *Collected Essays on Learning and Teaching* 5: 33–38.
Horton, R. 2010. Anew epoch for health professionals' education. *The Lancet* 376: 1875–1877.
Hunter, B., P.W. George, and G.C. Godbey. 2006. What does it mean to be globally competent? *Journal of Studies in International Education* 10 (3): 267–285.
ILO (International Labour Organization). 1999. *More than 140 million denied access to health care in Latin America and the Caribbean.* http://www.ilo.org/global/about-the-ilo/newsroom/news/WCMS_007961/lang--en/index.htm.
Jarvis-Selinger, S., D.D. Pratt, and G. Regehr. 2012. Competency is not enough: Integrating identity formation into the medical education discourse. *Academic Medicine* 87: 1185–1190.
Kennedy, M.D. 2015. Globalizing knowledge. In *Intellectuals, Universities, and Publics in Transformation.* Stanford: Stanford University Press.
Lingard, L. 2012. Rethinking competence in the context of teamwork. In *Culture and Politics of Health Care Work: Reconsidering Medical Education in the Twenty-First Century*, ed. B.D. Hodges and L. Lingard. Ithaca: ILR Press.
Ludmerer, K.M. 2015. Let me heal. In *The Opportunity to Preserve Excellence in American Medicine.* Oxford/New York: Oxford University Press.
Marmot, M. 2015. *The Health Gap. The Challenge of an Unequal World.* London: Bloomsbury.
Massey, D. 2004. Geographies of responsibility. *Geografiska Annaler. Series B. Human Geography* 86: 5–18.
Milgram, S. 1967. The small world problem. *Psychology Today* 1: 61–67.
Noddings, N., ed. 2005. *Educating Citizens for Global Awareness.* New York/London: Teachers College Press.
Nussbaum, M.C. 1997. Cultivating humanity. In *A Classical Defense of Reform in Liberal Education.* Cambridge, MA: Harvard University Press.
Nussbaum, M. 2002. Education for citizenship in an era of global connection. *Studies in Philosophy and Education* 21: 289–303.

OXFAM. 2014. *Working for the Few*. 178 Oxfam Briefing paper. http://www.oxfam.org/sites/www.oxfam.org/files/bp-working-for-few-political-capture-economic-inequality-200114-summ-en.pdf.

Parkin, D., J. Appleby, and A. Maynard. 2013. Economics; the biggest fraud ever perpetrated on the world? *The Lancet* 382: e11–e15.

Pegoraro, R., and G. Putoto. 2007. Findings from a European survey on current bioethics training activities in hospitals. *Medicine, Health Care, and Philosophy* 10 (1): 91–96.

Persad, G.C., L. Elder, L. Sedig, L. Flores, and E.J. Emanuel. 2008. The current status of medical school education in bioethics, health law, and health economics. *The Journal of Law, Medicine & Ethics* 36 (1): 89–94.

Piketti, T. 2014. *Capital in the Twenty-First Century*. Cambridge, MA/London: The Belknap Press of Harvard University Press.

Potter, V.R. 1971. *Bioethics. Bridge to the Future*. Englewood Cliffs: Prentice Hall.

———. 1988. *Global Bioethics. Building on the Leopold Legacy*. East Lansing: Michigan State University Press.

Quisumbing, L.R. 2005. Education for the world of work and citizenship: Towards sustainable future societies. *Prospects* 35 (3): 289–301.

Reimers, F. 2009. Enlightening globalization: An opportunity for continuing education. *Continuing Higher Education Review* 73: 32–45.

Reysen, S., and I. Katzarska-Miller. 2013. A model of global citizenship: Antecedents and outcomes. *International Journal of Psychology* 48 (5): 858–870.

Riffkin, R. 2014. In US, 67% Dissatisfied with Income, Wealth Distribution. *Gallup*, January 20, 2014. http://www.gallup.com/poll/166904/dissatisfied-income-wealth-distribution.aspx.

Rigolini, J. and R. Vakis. 2017. Four Facts About Poverty in Latin America you Probably Didn't Know. *The World Post*. http://www.huffingtonpost.com/jamele-rigolini/four-facts-about-poverty-_b_6820048.html.

Robinson, F. 2011. *The Ethics of Care. A Feminist Approach to Human Security*. Philadelphia: Temple University Press.

Schattle, H. 2008. *The Practices of Global Citizenship*. Lanham: Rowman & Littlefield Publishers.

Stiglitz, J.E. 2012. *The Price of Inequality. How Today's Divided Society Endangers our Future*. New York/London: W.W. Norton & Company.

Stirrat, G.M., C. Johnston, R. Gillon, and K. Boyd. 2010. Medical ethics and law for doctors of tomorrow: The 1998 consensus statement updated. *Journal of Medical Ethics* 36 (1): 55–60.

Stromquist, N.P., and K. Monkman, eds. 2000. *Globalization and Education. Integration and Contestation Across Cultures*. Lanham: Rowman & Littlefield Publishers.

Swiss Consulting Group. 2002. *Global Competence Report*. Zurich: Swiss Consulting Group.

Taylor, C. 1985. *Human Agency and Language: Philosophical Papers I*, 1985. Cambridge: Cambridge University Press.

Ten Have, H. 2008. Unesco's ethics education Programme. *Journal of Medical Ethics* 34 (1): 57–59.

Ten Have, H.A.M.J. 2014. Bioethics education. In *Handbook of Global Bioethics*, ed. H. ten Have and B. Gordijn, 447–466. Dordrecht: Springer.

Ten Have, H. 2015. Globalization of bioethics education. In *Bioethics Education in a Global Perspective*, ed. H. ten Have, 1–19. Dordrecht: Springer.

———. 2016. *Global Bioethics. An Introduction*. New York: Routledge.

Ten Have, H.A.M.J., and B. Gordijn. 2014. *Handbook of Global Bioethics*. Dordrecht: Springer Publishers, 4 volumes.

UNESCO. 2015. Rethinking education. In *Towards a Global Common Good*. Paris: UNESCO.

United Nations. 2003. *Report on the World Social Situation: Social Vulnerability: Sources and Challenges*. New York: United Nations Department of Economic and Social Affairs.

Wass, V. 2015. Opening our eyes to Global Health: A philosophy of universal values. *Perspectives on Medical Education* 4: 331–333.

Wass, V., and L. Southgate. 2017. Doctors without borders. *Academic Medicine* 92 (4): 441–443.

WHO. 2006. *World Health Statistics*. Geneva: WHO. http://www.who.int/whosis/whostat2006.pdf.
———. 2007. *World Health Report 2007 – A Safer Future: Global Public Health Security in the 21st Century*. Geneva: WHO.
Wintrup, J. 2015. The changing landscape of care: Does ethics education have a new role to play in health practice? *BMC Medical Ethics* 16: 22. https://doi.org/10.1186/s12910-015-0005-0.
World Hunger Education Service, *Hunger Notes*. 2016. *World hunger and poverty fats and statistics*. http://www.worldhunger.org/2015-world-hunger-and-poverty-facts-and-statistics.

Chapter 2
Global Bioethics and Global Education

Solomon Benatar

Abstract A new context for ethics and ethics education is evident in a rapidly changing world and our threatened planet. The current focus on considerations of inter-personal ethics within an anthropocentric perspective on life should be extended to embrace considerations of global and ecological ethics within an eco-centric perspective on global and planetary health. The pathway to understanding and adapting to this new context includes promoting shifts in life styles from selfish hyper-individualism and wasteful consumerism towards cautious use of limited resources within an increasingly interdependent world in which the equal moral worth of all and sustainability are valued. Critical scholarly approaches to global politics and to the global political economy could facilitate such change and encourage iterative interactive processes instead of seeking conclusive definitive 'scientific' solutions to all problems. Hopefully this shift in perspective could be achieved firstly through sensitization to new and increasingly challenging ethical dilemmas, and then by encouraging rational thinking and action based on global and ecological considerations rather than on false economic dogma and the distorted workings of a market civilization. Moving ahead with these activities must begin with promotion of education, learning and self-reflection to foster the widespread development of a global state of mind. Such a shift would require an expanded ethical discourse, with consideration of ethical dilemmas beyond human inter-personal relationships. These should include intra-institutional and inter-institutional relationships, as well as relational ethics between nations in a post-Westphalian world, and between humans and nature in an era now called the Anthropocene, to ensure survival on a planet undergoing entropy.

The adverse effects of intense competition could be reduced by encouraging greater cooperation, linking security issues to global health and social justice issues, and using inspirational narratives and examples of moral imagination/moral leadership. These educational programs should be initiated in schools and become mandatory for all first-year College and University students. Such an agenda is feasible

S. Benatar (✉)
University of Cape Town, Cape Town, South Africa

Dalla Lana School of Public Health, University of Toronto, Toronto, Canada
e-mail: Solomon.benatar@uct.ac.ca

H. ten Have (ed.), *Global Education in Bioethics*, Advancing Global Bioethics 10,
https://doi.org/10.1007/978-3-319-78984-2_2

given human ingenuity, determination, resilience and adequate resources, but will also require political and social will as its drivers.

2.1 An Evolving Global Context for Ethics, Bioethics and Ethics Education

The world in the first quarter of the twenty-first century is more different from, and changing more rapidly, than was anticipated 100 years ago – with profound implications for how we shall live in the future and for how we think we should live.

The pace of change was already evident when in 1958 Aldous Huxley looked back on his 1932 predictions for several centuries into the future (Huxley 1932; Huxley 1958). For example, he noted that in those countries with a tradition of democratic government, freedom was on the decline and the shift toward totalitarian forms of control was advanced far beyond what he had predicted. Commercial and political organizations had already developed new techniques for manipulating the thoughts and feelings of the masses, and population growth was beginning to outstrip food production in poor countries. In 1958 Galbraith also provided prescient insights regarding the implications of already evident rapid social change in his book 'The Affluent Society'(Galbraith 1958). These insights were largely ignored and the trend has continued almost exponentially (Franklin 1999; Halliday 2001).

It is notable that by the 1960s an impressive recovery had been achieved from the devastation caused by the 1918 Spanish flu epidemic that killed 50–100 million people (3–5% of the world's population), the Great Crash of 1929 and the two world wars that killed 70 million, maimed vast numbers of people and destroyed many cities. The moral and social victories achieved over the dark genocidal forces that had been unleashed by Hitler enabled the rebuilding and restructuring of social worlds within the supportive framework of Keynesian economic policies. Cities in England and Europe were re-built, with social services strengthened to sustain universal access to education, health-care, public transport, and other benefits that would make life more tolerable for many more people than in the pre-war liberal-capitalist era. Similar but less enduring effects were being achieved under the New Deal in the USA. Moreover, with considerable financial support from the USA, Japan was recovering from the devastating human and social impact of the atomic detonations. Colonizing nations were facilitating independence of their colonies and the world economy was expanding steadily.

Rapidly advancing medical knowledge included sophisticated new diagnostic techniques that allowed exquisite visualization of body structures and functions and effective medical advances, including many new medications, Intensive Care Unit facilities and organ transplantation were saving many lives. These expensive medical advances that could not be applied to all who needed them spawned many ethical dilemmas and diminishing trust in the competence of the medical profession to make complex ethical decisions associated with the fair allocation of access to such life-saving endeavors. Theologians and philosophers came out of their professional closets to analyze the ethics of medical practice and fueled the demand for enhanced patient decision-making power.

Further boosting of respect for individuals through the concomitant civil rights thrust, spurred the emergence of individual autonomy in the 'new medical ethics' (bioethics) in the 1970s (Johnson 1990; Walter & Klein 2003). Since then this field has grown to the point where today professional bioethics centers are well established in many countries and make valued contributions to complex ethical decisions in everyday practice. Encyclopedias describing many facets of advances in bioethics as a new field of human endeavor and the expanding ethical regulatory frameworks for medical research and practice, are further testimony to the growth of such activities (Reich 1995).

Between 1960 and 2010 the world's population increased along an exponential growth curve from just over 3 billion to almost 7 billion. Energy use also increased exponentially within what was becoming a 'market civilization' underpinned by the 'technological imperative'. During the twentieth century, wondrous scientific and technological progress, was associated with use of ten times the amount of energy than the total used in the preceding 1000 years.

Yet new ethical challenges were becoming prominent in association with the emergence of a range of interconnected global crises. Widening disparities in wealth and health, resurgence and global spread of old infectious diseases such as tuberculosis and malaria, rising drug resistance and the emergence of new infectious diseases (HIV, SARS, MERS, Chicungunya, Ebola and Zika) stimulated expansion of the bioethics discourse to include public health ethics (Thompson et al. 2013).

During the early twenty-first century even larger crises were seen to be looming – global economic crises, escalating social conflict, proliferation of fundamentalist groups, ethnic and religious conflict, anarchy and failure to reproduce caring social institutions (such as educational, health and other infra-structural services), ongoing population growth, wasteful consumption patterns by the privileged, severely deprived living conditions for billions and most significantly global warming, environmental degradation and escalating migration of people displaced by war or natural disasters. Militarism and militarization driven by the technological thrust of the industrialized world led to the (still present) threat of nuclear war, and contributed to world economic crises causing poverty and starvation (associated with competition for, and exploitation of, shrinking natural resources), with resulting environmental degradation and pollution (as the costs of protecting nature were not built into the consumerism associated with a market civilization).

It has become increasingly apparent that these escalating crises have profound effects on the future of life and on planetary sustainability in this Anthropocene epoch of the Holocene, in which human activity is depleting non-renewable planetary resources on which all life is so critically dependent, and threatening to overwhelm the widespread beneficial application of many scientific and medical advances (Potter 1971; Steffen et al. 2015).

The abuse of human rights has been an associated structural feature of this complex chain of events, despite the Universal Declaration of Human Rights and much work done to propagate the concept of human rights at a global level (Benatar

1998). It is sobering that Eric Hobsbawm, the eminent historian, noted that: "I have lived through most of the twentieth century without, I must add, suffering personal hardship. I remember it only as the most terrible century in Western history" (Hobsbawm 1995). In his reflections on the twentieth century Hobsbawm refuted the notion of this century as one of undiluted progress and instead described it as "an age of extremes." He labeled 1914–1950 as an age of catastrophe during which two world wars raged; 1950–1973 as a golden age in which war-torn countries in Europe were rebuilt and America prospered; and more importantly for the perspective of this chapter he perceived the period 1973–1991 as the landslide towards the unknown future in the third millennium. His prescience was surely remarkable.

While many may wish to avoid deep introspection about such a depressing scenario, critical thinkers acknowledge that scientific evidence of relentless entropy associated with what is called 'progress,' requires re-examination of a value system and processes that have led to our current predicament with its associated ethical implications for human behavior (McMichael 2014; Benatar 2011).

Most recently there has been an exacerbation of cultural conflicts with new and intense focus on racism and identity politics that have even further complicated and impeded cross-cultural dialogue and the quest for peaceful progress. This background and trajectory have shaped an evolving context for extension of our current focus on considerations of inter-personal ethics from an incomplete western and anthropocentric perspective on life, towards a more embracing global, planetary and ecological conception of health (Benatar et al. 2003, 2009; Benatar and Brock 2011).

2.2 The Pathway to Understanding and Adapting to This New Context

The pathway to adapting to this new context begins with understanding the complex historical trajectory of 'progress' from the Middle Ages through the Enlightenment into an epoch marking the potential for retrogression into another dark period (Tarnas 1991; Wright 2004; Chu and Karr 2017). Achieving such understanding will need deep introspection into ways of life that have come to be accepted as 'normal'. Paradigm shifts in thinking and action that could promote beneficial shifts in life styles will also be required. The goal should be to achieve progression *from* selfish hyper-individualism, wasteful consumerism, narrowly focused biomedical approaches to improving health, and exploitative extraction processes that undermine billions of lives, *towards* more cautious and shared use of limited resources within an increasingly interdependent world in which greater value could be accorded to the equal moral worth of all and to sustainability of a life supporting environment deep into the future (Benatar et al. 2009). Awareness of this new context for human flourishing could and should be the catalyst for developing an extended discourse on ethics.

2.3 From Individual Health to International Health and Global Health

Medical advances in improving the health of individuals including the new biology with personalized precision treatments and genetic advances, have become most widely applied in wealthy countries and the new bioethics has evolved in response to the need for morally justified application of medical advances in those societies. Wide disparities in wealth and health across the world with the associated challenges of achieving human well-being and security have also led to proliferation of international health activities to bring such biomedical advances to the poor. These well-intentioned, individualistic and biomedical contributions to the modern medical care of individuals, together with public health measures supported by charitable and philanthropic endeavors (often through public-private partnerships) have indeed become (popularly and somewhat myopically) accepted as *the priority* for improving health in poor countries across the globe (Daar and Singer 2011), with neglect of the social and societal determinants of health (Birn et al. 2017).

The historical antecedents of current international health activities began with colonial medicine in the eighteenth and nineteenth centuries (Birn 2009). Tropical disease medicine followed in the early twentieth century when advances in medicine facilitated control of infectious diseases. International health activities expanded and changed in the mid twentieth century when infectious diseases were largely controlled in wealthy countries and several co-operative endeavors were implemented to promote and improve health in low and middle–income countries. The global program to control HIV/AIDS exemplifies the desired success from such projects.

However, it has become clear that while such efforts are necessary on an increasingly wide scale, they will not be sufficient, as they do not address the root causes of poverty that enable the emergence and spread of new infectious diseases, antibiotic resistance and conflict that, inter alia, affect health adversely (Garrett 1994; Garrett 2000). With increasing globalization of many social and economic activities international health work has been renamed ‘global health,’ but without any change in the nature of such activities that remain largely focused on health within a biomedical paradigm. A distinction must be made here between the abovementioned international health/global health activities and a more comprehensive understanding of what global health truly means in an era in which the challenges are greater than ever.

It is arguable that this new understanding of global health (more accurately articulated as health of all life on our planet), in the much broader context of an interdependent world characterized by threats to the inter-linked health of humans, non-human species and planetary ecology and resilience, calls for a much more comprehensive understanding of the forces that improve or damage health in the broadest sense. This concept of *global health* thus goes well beyond the traditional concept of *international* health that focused on bilateral interactions between wealthy and poor countries, with biomedicine as the driving force. It must reach beyond the rich-poor dichotomy and geographic borders to include the socio-political-economic forces that separate the powerful, free, privileged population from those who are powerless, lack freedom and are marginalized (Markle et al. 2014). Acknowledging the accuracy and legitimacy of this description of global health, sets the scene for developing enhanced sensitivity to ethical dilemmas within major institutions and at the level of international relations.

Once global health is defined and acknowledged in this broader planetary and ecological sense, the next step is to examine the state of human health in the world and critically analyze how this has come about. Much factual information is available including data about global disparities and the deteriorating state of planetary resilience (Benatar and Brock 2011). Rather than attempting to document this here it will suffice to briefly and succinctly illustrate the state of human health globally by reference to such variations between the poorest and wealthiest people as life expectancy at birth from 35 years to 85 years; under 5 year mortality from 20:1000 to 170:1000 live births; maternal mortality from ranging from 7 deaths per 100,000 pregnancies in Canada, to 134 in South Africa, 789 in Southern Sudan and 1360 in Sierra Leone; and annual per capita expenditure on health care from about $30 to over $8000.

The health and resilience of our planet is similarly reflected in the facts that: a considerable proportion of the world's tropical and temperate forests have been destroyed and the rate of deforestation in the tropics continues at about an acre a second; much of the world's wetlands, and the diversity of life and ecological services they provide have disappeared; a majority of large predatory fish are gone and about 75% of marine fisheries are now overfished or fished to capacity; widespread soil depletion has resulted in continued agricultural production being possible only with massive inputs of fertilizer and pesticides, as well as artificial irrigation from rapidly depleting underground aquifers; dozens of persistent toxic chemicals have been measured in many sites; the consequences of climate change are already being felt, in some cases several decades before they were forecast to occur (Speth 2008).

2.4 Causal Underpinnings

Although the above data is widely available, the social and political causal underpinnings of such developments and their long-term implications are largely ignored or denied (Stiglitz 2015). Reasons for this denial in part relate to our assumption that the values and theories (such as current economic theories) underlying scientific, technological and medical advances, together with spectacular economic growth are adequate for ongoing progress towards improved health for humans and our ecology.

Yet honest analysis and critical examination of the causal pathways to the evolution of global health disparities, uncover global economic, political and business influences that together with unrealistic expectations and entitlements (denial of the limits of life and medicine) and dominance of the technological imperative have undermined population health and distorted health care 'systems', making them dysfunctional and unsustainable. Lack of leadership in global health governance and failure to address the upstream causes of poverty that account for 75% of premature deaths and destruction of the ecological milieu, which underpins sustainable health, are symptomatic of these shortcomings (Gill and Benatar 2016).

The trajectory towards such dysfunction in the face of spectacular progress can be seen to be at least in part related to a shift from an ideology that valued both community and individuals within communities (for example the Keynesian economic model that enabled post WW II reconstruction of the global economy and social structures supportive of education and health care as common public goods) towards a neoliberal ideology that distorts individualism to hyper-individualism, placing the highest values on individuals with little regard for communities.

2.5 The Currently Dominant Belief System

The dominant post-Enlightenment neoliberal belief system and its frames for global thinking have been caricatured as emphasizing individualism, human rights, freedom, philanthropy and a minimally restrained market economy. The neoliberal market civilization model corresponding to this belief system is central to the economic, social and cultural development of contemporary capitalism. Amongst other things, it involves structural acceleration in processes within the world economy that together adversely impinge on the health of populations and sustainability of the biosphere. In their most extreme expression, as noted elsewhere these processes, undertaken at the expense of conditions for healthy living, and for delivery of healthcare, include:

> (1) intensification of the exploitation of human beings, social processes and nature for purposes of profit; (2) incremental dispossession of communities of their basic and local means of subsistence and livelihood; (3) acceleration in the turnover time of the production and sale of commodities to generate quicker accumulation of profits for firms and investors; and (4) restructuring or privatization of previously public institutions and public goods, including provisions for healthcare and education. As the latter become increasingly subject to market forces and values, they are treated like commodities that can be simply bought and sold for reasons of profit. (Gill and Benatar 2016, p. 350).

Neoliberalism is also characterized by a very narrow conception of human rights as a dominating moral language, with inadequate attention to the ethics of responsibilities, duties, virtues, needs and solidarity. This concept of rights is excessively focused on freedom to do what we want to do, to the exclusion of the right of freedom from oppression and denial of the basic means of subsistence. Short-term self-interest eclipses longer-term interests, with consequent erosion of a sense of community and moral imagination. Unlimited wants within a market and consumerist ideology pervade all aspects of life and lead to living beyond our means, exponential consumption and the dominance of corporate rights and goals over the needs of individual people.

Moreover, faith in science and technology (as *the* solutions to all problems), and the valuing of new knowledge over old knowledge, and of knowledge over wisdom in the application of knowledge, enables disjunction between economic growth, advances in science, technology and medical care, and our ability to use these advances to improve health more widely.

2.6 Moving Ahead with Activities to Improve Health Locally and Globally

All the above issues should be included in education about global health, as deeper knowledge of these causal pathways could help illuminate what needs to be rectified in the quest for understanding our current predicament and answering questions about what ought to be done and with what justification. It is surely correct that in the context of the serious health implications of climate change and environmental degradation, new paradigms of thinking and action are required to ensure meaningful progress in the health of whole populations and the sustainability of life on our planet. Consensus will not be easy to achieve, as how we view ourselves, the world in which we live, the kind of future world to which we aspire, and what may be the most appropriate research agenda for the pursuit of such goals, all depend on our *belief system*.

The ability to think critically should be extended to include identifying and pursuing a more comprehensive value system within which normative responses could be developed and widely accepted. The promotion of rational discourse on the pros and cons of varied approaches would follow, taking into consideration global political and ecological issues such as interdependence, solidarity and long-term sustainability. This is a much broader and deeper agenda than that associated predominantly with a biomedical conception of health.

So, moving ahead with activities to improve health in this broader sense must begin with promotion of education, learning and self-reflection to foster widespread development of a *global state of mind*, i.e. thinking of ourselves as intimately connected with and dependent on all forms of life on our planet, understood as a self-renewing complex system with limited resilience. Such a shift in outlook would require inter-alia, knowledge about the epidemiology of health and disease globally, some understanding of causal pathways and developing an expanded ethical discourse. To facilitate survival on a planet undergoing entropy, consideration of ethical dilemmas should be extended beyond the micro-level (inter-species and inter-personal relationships) to the meso level (intra-institutional and inter-institutional relationships) and to the macro level (international relationships and relationships with nature) in an era now known as the Anthropocene.

2.7 Global Health Ethics

Global health and global health ethics education should be informed by multiple disciplines, inclusive of health, science, philosophy, anthropology, psychology, sociology, history, geography, economics, religion, and law to achieve the well-being of all people within the worldwide community, regardless of differences based on race, ethnicity, political affiliation, economics, culture, education, language, gender, age, or religion (Markle et al. 2014). An extended discourse and public debate propagated through such a multi-disciplinary approach to ethics in education and daily life, could enhance the realization that health, human rights, economic opportunities, good governance, peace and development are all intimately linked within a complex, interdependent world (Stapleton et al. 2014).

The moral challenge of the twenty-first century is to explore these links, to understand their implications and to develop processes that could harness economic growth to human development, narrow global disparities in health and promote peaceful global coexistence (Benatar 2005). A global agenda must extend beyond the rhetoric of universal human rights to include greater attention to duties, social justice and interdependence. Health and ethics provide a framework within which such an agenda could be developed and promoted across borders and cultures.

A new multi and trans-disciplinary field of bioethics, when expanded in scope to embrace widely shared foundational values, could make a valuable contribution to improving global health (Benatar et al. 2003). A framework for action that combines understanding of global interdependence with enlightened long-term self-interest has the potential to produce a broad spectrum of beneficial outcomes, especially in global health. Our vision, described elsewhere, offers a way forward for global health reform through several transformational approaches, and is briefly summarized here.

Developing a global state of mind about the world, and our place in it, it is perhaps the most crucial element in the development of an ethic for global health. In arguing that it is both desirable and necessary to develop a global mindset in health ethics, we suggest that this change need not be based merely on altruism. Promoting *long-term self-interest* through *solidarity* with others is also essential, if we acknowledge that lives across the world are inextricably interlinked by globalizing forces that, through rapid communication and mobility powerfully shape health and well-being. Striking a balance between optimism and pessimism about globalization, solidarity and progress will require a platform for dialogue among stakeholders, and a space where people can share different views about globalization.

A broader conception of bioethics also offers a basis for space to optimize the benefits of globalization, and reduce its adverse impacts. Our vision for promoting an ethic for global health also features *development of the human capacity* of a greater proportion of the world's population to achieve their potential, and a commitment to a broader discourse on ethics propagated through centers regionally and globally networked in growing and supportive North-South partnerships (Kickbusch 2000). We aspire to the hope that the adverse effects of intense competition could perhaps be reduced by encouraging greater cooperation, linking security issues to global health and social justice issues (Gill and Benatar 2017), and using inspirational narratives and examples of moral imagination and moral leadership (Lowry et al. 2009; Kleinman 2006).

Even more difficult would be to move the discourse toward a bioethical-political agenda to drive change towards achieving widespread access to such public goods as health care, education, basic subsistence needs and employment and the associated collective action for more equitable financing (Kaul et al. 1999). Constructing new ways of achieving economic redistribution is arguably the key to resolving many global problems (Soros 2002). Good governance could enhance the optimum distribution and use of public goods. Such an agenda is feasible given human ingenuity, determination, resilience and resources, but will also require political and social will as its drivers (Gill and Benatar 2016).

2.8 Teaching Global Health and Global Health Ethics

An accumulating number of texts addressing one or more aspects of international health, global health, global health ethics and global ethics, provide rich resources for education about global health and relevant ethical considerations. Catharine Myser's edited volume deals with such issues as the perils and pitfalls of globalizing western bioethics, the range of socio-cultural interests and forces shaping bioethics around the world, bioethics as a means for negotiating social, regional and national identity, bioethics as nation-building and bioethics as a battleground for religious and political culture wars (Myser 2011). Ronald Green and colleagues focus their attention on the normative bases for bioethics, various aspects of global research ethics, biomedical and bioethics collaborations, training professionals, and such specific subjects as euthanasia and physician assisted suicide, global bioethics and religion and public global bioethics consultations (Green et al. 2008).

Ruth Macklin addresses issues in research ethics at the intersection of global health, justice and human rights (Macklin 2012). Joseph Millum and Ezekiel Emanuel's edited volume is similarly focused on global aspects of research ethics (Millum and Emmanuel 2012). The topic of ethics and infectious diseases is dealt with by Margaret Battin and colleagues while issues of social justice lie at the heart of Lisa Eckenwiler's book (Battin et al. 2009; Eckenwiler 2012). Issues pertinent to global social justice and health are also covered by Heather Widdows and Nicola Smith while John Coggon and Swati Gola seek to explicate ethical, political and regulatory challenges in their book on global health (Widdows and Smith 2011; Coggon and Gola 2013). Paul Farmer and colleagues add anthropological, biosocial and narrative perspectives (Farmer et al. 2013). Two other major textbooks focus on a wider range of global health issues (McKracken and Phillips 2012; Birn et al. 2017). Two other recent books deal more comprehensively with global health and global health ethics, and yet another two with Global Ethics and Global Bioethics (Benatar and Brock 2011; Pinto and Upshur 2013; Singer 2002; Ten Have 2016).

A perennial question is whether teaching ethics can improve ethical behavior. It is arguable that ethics education should not be expected to directly improve ethical action. Hopefully sensitization to the ethical dimensions of life and to how to reason about ethical dilemmas could enhance the ability of members of an intelligent species to modify their individual and collective behavior. In the new *International Journal of Ethics Education*, Diego Gracia has defined the goal of education as the promotion of intellectual, moral and human skills as well as the character of human beings (Gracia 2016).

Donald Cole and colleagues have argued that a primary goal of teaching global health and global health ethics is to enhance competence in ethical reasoning. The three educational approaches they recommend for fostering appropriate attitudes and practices are experiential learning (including systematic reflection), critical pedagogies (that question unjust power, for example within institutions) and transformative learning (education for change), examples of all of which can be found in several web-based curricula. They note that global health ethics teaching can take place in a variety of informal and formal settings with collaboration across disciplines to enable deeper understanding of alternative (cultural) conceptualizations of health. Finding 'teachable moments' provides multiple and diverse opportunities to enable institutions to support the global health ethics teaching endeavor. Their recommendations are illustrated through case studies (Cole et al. 2013).

James Dwyer has integrated his teaching of global health ethics into his ethics courses. Initially he emphasized disparities in health around the world and discussed theories of justice to focus on how unjust disparities could in theory be addressed. When he found that students were somewhat disinterested in the details of such theories he moved towards considerations of possible reactions and responsiveness of those who witness such disparities during global health electives, and how creative responses could be encouraged (Dwyer 2011; Dwyer 2015; Dwyer 2013). Two issues with which he has grappled in his teaching were the questions of engagement and hope. Regarding engagement, he argues that teachers should not remain totally detached from the world in which they live. While he views some detachment as permissible to enable reflection and theoretical understanding, he argues that this is insufficient and must be combined with some degree of engagement with activities to improve health. With respect to hope, he finds it hard to avoid pessimism in the face of the magnitude of the challenge to attempt to change deeply entrenched social structures. Yet he argues for hope that can encourage striving, resilience and readjustment to counteract the passivity, resignation and cynicism associated with pessimism. This tension is one with which he and surely all who think deeply about such issues need to grapple continuously (Dwyer 2011).

In her exploration of the value of case study pedagogy for a full-semester graduate course in global health ethics at an American university, Kearsley Stewart addresses some of the pedagogical challenges of teaching global health ethics through innovative use of case study methodology. She builds on the work of James Dwyer who, after many years of teaching bioethical principles and theories of justice, no longer emphasizes the differences between different theories of justice but instead focuses student learning on 'responsibility and responsiveness' to facilitate their wrestling with the breadth and depth of moral concern in the realm of global health. As Dwyer's approach directly links to the many conundrums Stewart faces in her graduate classroom, she has developed a case study method to transform his insights into a class that engaged some of the self-reflection central to all three of Cole et al.'s recommended pedagogies. She describes the rationale for, and the implementation of, the prospective case study assignment as a first step toward developing pedagogical innovations appropriate for teaching global health ethics at the master's level in the United States. In addition to standard didactic approaches to ethics education this introduces active learning, group work, peer instruction, and multiple opportunities for students to engage in self-reflection and become aware of their own learning (Stewart 2015).

2.9 Conclusions

It is undisputed that we live in an era of great threat to sustainable human and planetary well-being. How long we could continue, without changing our entrenched habits and before destroying our civilization, has been estimated but is uncertain. It remains an open question whether we can do more to prevent tragic future events and reverse current trends. If there is any hope, this must surely begin with ideas, continue with education and translate into political and economic action. This is the challenge for global bioethics and ethics education.

References

Battin, M.P., L.P. Francis, J.A. Jacobson, and C.B. Smith. 2009. *The Patient as Victim and Vector: Ethics and Infectious Disease*. Oxford: Oxford University Press.

Benatar, S.R. 1998. Global disparities in health and human rights: A critical commentary. *American Journal of Public Health* 88 (2): 295–300.

———. 2005. Moral imagination: The missing component in global health. *Public Library of Science Medicine* 2 (12): e400.

Benatar, S R. 2011. Global Leadership, Ethics and Global Health: The search for new paradigms. In *The Global Crisis & the Crisis of Global Leadership*, ed. S. Gill, 127–143. Cambridge: Cambridge University Press.

Benatar, S., and G. Brock, eds. 2011. *Global Health & Global Health Ethics*. Cambridge: Cambridge University Press.

Benatar, S.R., A. Daar, and P.A. Singer. 2003. Global health ethics: The rationale for mutual caring. *International Affairs* 79 (1): 107–138.

Benatar, S R, S. Gill, and I. Bakker I. 2009. Making progress in global health: The need for a new paradigm. *International Affairs* 85(2): 347–371.

Birn, A. 2009. The stages of international (global) health: Histories of success or successes of history? *Global Public Health* 4 (1): 50–68.

Birn, A., Y. Pillay, and T.H. Holtz. 2017. *Textbook of Global Health*. 4th ed. Oxford: Oxford University Press.

Chu, E.W., and J.R. Karr. 2017. Environmental impact: Concept, consequences, measurement. In *Reference Module in Life Sciences*. Elsevier Press. https://doi.org/10.1016/B978-0-12-809633-8.02380-3. isbn:978-0-12-809633-8.

Coggon, J., and S. Gola. 2013. *Global Health and International Community*. London: Bloomsbury.

Cole, D, L. Hanson, K.D. Rouleau, K. Pottie, and N. Arya. 2013. Teaching global health ethics. In *An Introduction to Global Health Ethics,* eds. A.D. Pinto and R.E.G. Upshur, 148-158. London/New York, Routledge.

Daar, A., and P. Singer. 2011. *The Grandest Challenges*. Canada: Doubleday.

Dwyer, J. 2011. Teaching global health ethics. In *Global Health and Global Health Ethics*, ed. S. Benatar and G. Brock, 319–328. Cambridge: Cambridge University Press.

———. 2013. On flying to ethics conferences: Climate change and moral responsiveness. *International Journal of Feminist Approaches to Bioethics* 6 (1): 1–18.

———. 2015. On taking responsibility for undocumented migrants. *Public Health Ethics* 8 (2): 139–147.

Eckenwiler, L.A. 2012. *Long-Term Care, Globalization & Justice*. Baltimore: Johns Hopkins University Press.

Farmer, P., J.Y. Kim, A. Kleinman, and M. Basilico. 2013. *Reimagining Global Health: An Introduction*. Berkeley: University of California Press.

Franklin, U.M. 1999. *The Real World of Technology*. CBC Massey Lectures, Rev Ed., Toronto: Anansi Press.

Galbraith, J.K. 1958. *The Affluent Society*. Harmondsworth: England.

Garrett, L. 1994. *The Coming Plague: Newly Emerging Diseases in a World out of Balance*. New York: Farrar, Straus & Giroux.

———. 2000. *Betrayal of Trust: The Collapse of Global Public Health*. New York: Hyperion.

Gill, S., and S.R. Benatar. 2016. Global Health governance and global Power: A critical commentary on the Lancet University of Oslo Commission Report. *International Journal of Health Services* 46: 346. https://doi.org/10.1177/0020731416631734.

———. 2017. History, structure and agency in global health governance: Comment on "global health governance challenges 2016 – Are we ready?". *International Journal of Health Policy and Management* 6 (4): 237–241.

Gracia, D. 2016. The mission of ethics teaching for the future. *International Journal of Ethics Education* 1: 7–13.

Green, R.M., A. Donovan, and S.A. Jauss. 2008. *Global Bioethics: Issues of Conscience for the 21st Century*. Oxford: Clarendon Press.

Halliday, F. 2001. *The World at 2000*. New York: Palgrave.

Hobsbawm, E. 1995. *The Age of Extremes: A History of the World. 1914–1991*. New York: Pantheon Books.

Huxley, A. 1932. *Brave New World*. New York: Harper.

———. 1958. *Brave New World Revisited*. New York: Harper Perennial.

Johnson, A.R. 1990. *The New Medicine and the Old Ethics*. Cambridge MA: Harvard University Press.

Kaul, I., I. Grunberg, and M.A. Stern, eds. 1999. *Global Public Goods: International Cooperation in the 21st Century*. New York: United Nations Development Programme.

Kickbusch, I. 2000. The development of international health policies: Accountability intact? *Social Science and Medicine* 51: 979–989.

Kleinman, A. 2006. *Living a moral life amidst uncertainty and danger*. Oxford: Oxford University Press.

Lowry, C., et al. 2009. Two models of global health ethics. *Public Health Ethics* 2 (3): 276–284.

Macklin, R. 2012. *Ethics in Global Health: Research Policy and Practice*. Oxford: Oxford University Press.

Markle, W.H., W.H. Fisher, and R.A. Smego. 2014. *Understanding Global Health*, 2nd ed. Chapter 21. London: Mc Graw Hill. http://accessmedicine.mhmedical.com/content.aspx?bookid=710&sectionid=46796922.

McKracken, K., and D.R. Phillips. 2012. *Global Health: An Introduction to Current & Future Trends*. London: Routledge.
McMichael, A. 2014. Population health in the Anthropocene: Gains, losses and emerging trends. *The Anthropocene Review* 1 (1): 44–56.
Millum, J., and E.J. Emanuel. 2012. *Global Justice and Bioethics*. Oxford: Oxford University Press.
Myser, C., ed. 2011. *Bioethics Around the Globe*. Oxford: Oxford University Press.
Pinto, A.D., and R.E.G. Upshur, eds. 2013. *Introduction to Global Health Ethics*. Oxford: Routledge.
Potter, V.R. 1971. *Bioethics: Bridge to the Future*. New Jersey: Prentice-Hall.
Reich, W., ed. 1995. *Encyclopedia of Bioethics*. New York: Simon and Schuster.
Singer, P. 2002. *One World: The Ethics of Globalization*. New Haven: Yale University Press.
Soros, G. 2002. *On Globalisation*. New York: Public Affairs.
Speth, J.G. 2008. *The Bridge at the End of the World*. New Haven: Yale University Press.
Stapleton, G., et al. 2014. Global health ethics: An introduction to prominent theories and relevant topics. *Global Health Action* 7 (23579): 23569.
Steffen, W., K. Richardson, J. Rockstrom, et al. 2015. Planetary boundaries: Guiding human development on a changing planet. *Science* 347 (6223.):1259855 1–10.
Stewart, K.A. 2015. Teaching corner: The prospective case study: A pedagogical innovation for teaching global health ethics. *Journal of Bioethical Inquiry* 21 (10): 57–61.
Stiglitz, J. 2015. *The Great Divide*. London: Allen Lane Penguin.
Tarnas, R. 1991. The Passion of the Western Mind. USA, Crown.
Ten Have, H. 2016. *Global Bioethics: An Introduction*. Oxford: Routledge.
Thompson, A, S.A. Nixon, R. Upshur, A. Robertson, S.R. Benatar, and A. Daar. 2013. Public Health Ethics. In Public Health Law & Policy in Canada, 3rd ed. T.M. Bailey, T. Caulfield, and N.M. Ries, eds, 37–57. Lexis Nexis Butterworths, Ontariio.
Walter, J.K., and E.P. Klein, eds. 2003. *The Story of Bioethics*. Washington, DC: Georgetown University Press.
Widdows, H., and N.J. Smith. 2011. *Global Social Justice*. London: Routledge.
Wright, R. 2004. *A Short History of Progress*, CBC Massey Lectures. Toronto: Anansi Press.

Part II
Goals and Challenges of Global Ethics Education

Chapter 3
Goals in Global Ethics Education

Volnei Garrafa and Thiago Rocha da Cunha

Abstract The teaching of global ethics can be developed from several points of view. The goals of this education should be to seek respect for plurality and construct a fairer, more equal and more supportive world. The reflections put forward in the present chapter are not only based on the theoretical foundations of ethics and bioethics, but also especially on the geopolitical *locus* of where the authors live and work, which is the southern hemisphere. The objective of using this artifice was to prevent local negative effects that an intendedly globalized project for teaching ethics might cause. The aim was also to indicate the conceptual bases and practices from which legitimate goals can be defined for this initiative, which can be applied for use in different parts of the world without unilaterality, interpositions, impositions or manipulations. Because of the difficulty in defining goals for teaching global ethics, the present analysis took the reference point of two real themes of the contemporary world: moral imperialism and the point of view of coloniality. Overcoming the distortions and inequities of moral imperialism that are now perceived in the relationships established among different individuals, cultures, countries and regions of the world, combined with coping with situations of coloniality – whether occasional or global – is part of a macro-ethical context that poses an obstacle to construction of non-discriminatory, equal and inclusive global ethics. Considering the barriers against construction of consensuses in this field, the present text proposes the use of education in global ethics as a tool that is applied for opposing situations of moral imperialism and coloniality. Finally, instead of proposing abstract goals that are set aprioristically, this chapter indicates the paradigm of corporeity as a universal marker for what should be sought morally (pleasure and health) and what should be avoided (pain and disease) in education

V. Garrafa (✉)
UNESCO Cathedra in Bioethics - International Center for Bioethics and Humanities, Brasília, Brazil

University of Brasília (UnB), Brasília, Brazil

T. R. da Cunha
Pontifical Catholic University of Paraná (PUC/PR), Curitiba, Brazil

Center of Studies on Bioethics and Diplomacy in Health - Fiocruz, Brasília, Brazil

H. ten Have (ed.), *Global Education in Bioethics*, Advancing Global Bioethics 10,
https://doi.org/10.1007/978-3-319-78984-2_3

programs regarding global ethics. Thus, this chapter provides a minimal axiological basis upon which it will be possible to establish goals, thereby avoiding educational practices in the field of ethics that will become yet another factual element serving the maintenance of unilateral and hegemonic interests and an unjust global order.

3.1 Introduction

The construction of theory and practice for teaching global ethics can be developed using various reference points. In fact, the objectives and goals of this field of education can only be adequately defined through characterizing different points of view and interpretations on the theme. Thus, the present chapter will begin by discussing what is essentially understood today by the expression "Global Ethics", which provides epistemological support for this chapter.

Global ethics addresses moral conflicts from which the causes, consequences and resolutions can be better evaluated based on a point of view that considers both global and local aspects of human actions. The broad range of areas of interest for global ethics includes ethical controversies relating to complex contemporary themes such as globalization (Singer 2002), human rights (Pogge 2008), global health (Benatar and Brock 2011), teaching and education (ten Have 2015), violence and war (Heinze 2016), climate change (Arnold. 2014), among other conflicts of eminently global nature.

In turn, "global bioethics" is a theoretical and normative expression of applied ethics that was consolidated at the turn of the millennium, especially after retrieval of the thoughts of Van Rensselaer Potter (1971), which were the precursor of this thread of bioethics. Potter was an American researcher who, at the beginning of the 1970s, proposed a set of goals for education in global ethics, with strong defense of the need for educational programs applied at a worldwide level that should aim "not only to enrich individual lives but to prolong the survival of the human species in an acceptable form of society" (Potter 1971, p. 87).

Bioethics has been consolidated from both a clinical perspective (Jonsen 2003) and principlist perspective (Beauchamp and Childress 2013), which differ from Potter's initial proposal. However, from the beginning of the 1990s onwards, a slow process towards expansion of this field beyond the traditional biomedical approach could be seen. The first criticisms regarding the application of American bioethics beyond Anglo-Saxon culture began in Europe (Holm 1995). New perspectives were then proposed for a normative foundation for bioethics that would be more appropriate for European morality (Rendtorff and Kemp 2000). During the 1990s, Latin America also went through a process of critical revision of the influences of American bioethics over local production, with proposals for some autochthonous approaches for this discipline, mostly regarding collective and social themes relating to public health (Garrafa 1995; Schramm 1996; Fortes 1998) and other collective conflicts of sociopolitical nature (Mainetti 2010).

In the context of American production, an expansion towards a global approach for bioethics also started to become necessary by the end of the 1990s. This was mainly due to having to deal with new ethical conflicts that originated from the process of internationalizing clinical trials involving human beings and the beginning of studies on the ethical implications of global health (Angell 1997; Benatar 1998; Macklin 1998).

The Universal Declaration on Bioethics and Human Rights, published by UNESCO in 2005, brought in an important normative and institutional contribution towards a worldwide approach to bioethics and consolidated the retrieval of the global perspective that had been present since the inception of the field. This document dealt directly with the theme of education on global ethics, highlighting the need to establish information, training and education processes at an international level, with the goal of "achieving a better understanding of the ethical implications of scientific and technological developments, in particular for young people" (UNESCO 2005).

Through considering this historical context of development and broadening of the concept of global bioethics and its legitimacy, the present chapter indicates some assumptions for defining the goals of global ethics education. Use of bioethics reference points to analyze the teaching of global ethics can be justified not only because global bioethics and global ethics itself are both contemporary correlated manifestations of applied ethics, but also because of the impossibility of creating some type of border between these interdisciplinary and overlapping fields.

This overlap between the two fields explains, for example, why for some authors bioethics is one of the topics that should be addressed by global ethics (Moellendorf and Widdows 2014), while for other authors global ethics is one of the many approaches that can be considered in developing bioethics (ten Have 2016). Even Potter had already characterized this overlap, highlighting that his understanding was that global bioethics should be comprehended as an expression of "global ethics monitored by biological knowledge, with a multidisciplinary focus" (Potter and Potter 1995, p. 186).

One issue that cannot be overlooked in the context of focusing on global ethics education relates to the interpretation that is usually given to its meaning. This is a construction that is directed preferentially to the individual sphere of people/professionals acting within various fields of knowledge and work. However, the academic proposal that is developed here was structured and directed preferentially towards broader interests of public, collective, geographic-regional and worldwide nature.

The following topics will initially address some of the challenges in defining goals for global ethics education, followed by a discussion of two uncomfortable yet real subjects in today's world: moral imperialism and coloniality. These subjects need to be comprehended and taken into account in order to understand and reach transformative objectives within this context. As will be discussed later on, supplanting the distortions and inequities caused by the moral imperialism that is currently observed in relationships that are established between different individuals, cultures, countries and regions of the world, combined also with coping with situations of coloniality (whether occasional or global) is part of a larger context of macro-ethics, in which construction of non-discriminatory, equal, inclusive and fairer global ethics is sought.

Thus, based on discussion of the challenges relating to construction of a global consensus, and taking these two topics as reference points, the present chapter will use global ethics education as a tool for opposition to the situations of moral imperialism and coloniality that are commonly observed in the world today. Lastly, an alternative basis using the paradigm of human corporeity is proposed, from which the goals of global ethics education can be solidly constructed in opposition to any form of domination.

3.2 Challenges Relating to Reaching a Global Ethics Consensus

Is it possible to establish a global educational program for applied ethics? What would be the normative basis used for this proposal? How can respect towards moral pluralism be guaranteed while, at the same time, agreeing on a common ethical approach for different communities around the world? How can it be avoided that a global education program might become an instrument for imposing a given culture or a given set of interests on other people, thereby emphasizing the interests of the strongest over those of the weakest?

It is not intended that the present chapter will provide final answers to these questions, which are at the center of deep divergences in various fields of knowledge (including philosophy, law and social sciences, as well as global ethics itself, among others). Instead, some possible answers for this complex context will be analyzed here. Thus, the authors' reflections in the present chapter are not based solely on the theoretical foundations of bioethics but are based especially on the geopolitical *locus* of where we live and work: the southern hemisphere. In a previous study focusing on vulnerability (Cunha and Garrafa 2016), we analyzed bioethics production published in different regions of the world and demonstrated that there are important theoretical and normative differences regarding the geographic locations of these productions. The conclusion of that study was that these differences should not be overlooked or minimized. Instead, they should be understood, valued and taken into consideration in a project in which a legitimate global basis for bioethics was sought. The conclusion that we reached in that study forms the basic premise that we used to develop these reflections on the viability and legitimacy of the definitions for global goals for global ethics education. In this sense, the objective of this approach was not only to seek possible conditions that could minimize any negative local effects that a globalized ethics education project might cause, but also particularly to indicate the bases upon which legitimate objectives and goals can be defined for this initiative, such that it might be applicable in different parts of the world without unilaterality, intrusions, impositions or manipulations.

Within the scope of bioethics, the problem of defining common universal morality has already been posed by several authors. Engelhardt (2006) provided one of the bluntest negative responses to this inquiry. In his work *Global Bioethics: the collapse of consensus,* he took the view that we live on a diverse planet, in cultural, religious, political and moral terms, and that these differences frequently become irreparably opposed or unintelligible. Engelhardt considered that not even modern rational deliberation – based on the tradition of the European Enlightenment – could provide common grounds for a universal moral consensus involving moral communities that are profoundly extraneous to each other, considering that in some cases the agents diverge regarding the basic premises and rules of evidence that are important for moral deliberation.

According to Engelhardt, in a world in which it is desired to remain secular, peaceful and plural, there is only one way to ensure a legitimate agreement between moral strangers: contractual procedures for permission and consent. In his view, this contractualist and procedural approach towards ethics allows moral strangers to establish rules of coexistence in which substantive values (regarding what is good, well or correct, for example) are not debated. This would allow relationships in which the diverging values of different moral communities would be preserved or modified legitimately.

Engelhardt made it clear that the plane of global relationships is structured on the free market, which provides ideal objective conditions for establishment of contractual relationships between moral strangers. This led Engelhardt to criticize any moral imposition that assumed intervention from either the state or intergovernmental global bodies, whether based on universal morality built on human rights (as proposed by the United Nations, UN) or on any other substantive universal concept of morality (as defended by major religions).

Besides Engelhardt, other authors in the field of bioethics have also posed the problem of the possibility of universal moral agreements based on the western philosophical tradition. However, some of these authors have provided alternatives differing from or even opposite to those put forward by Engelhardt. In Latin America for example, for almost two decades a group of bioethicists has posed the problem of the negative effects of vertical and acritical adoption of universal bioethics approaches. Thus, by the end of the 1990s, the legitimacy of a principlist approach of universal bioethics was already being discussed in this region. A linguistic metaphor was established, in which it was suggested that this field of applied ethics should be understood as a language composed of different dialects and idiolects, which should all be respected regarding their particularities and should be correlated in a non-hierarchical manner (Garrafa et al. 1999).

Since then, we have sought to propose a global approach towards bioethics based on critical consideration of the process of globalization and the power relationships that are involved in defining the world between central and peripheral regions (Garrafa 2000). Within the scope of what was initially called "hard bioethics" and soon afterwards "intervention bioethics", it was sought to draw attention to the need for the bioethics agenda to start to incorporate persistent ethical conflicts that afflict most of the world's population, including matters that had been neglected, such as poverty, social exclusion, discrimination and other social issues that traditional bioethics did not consider at that time (Garrafa and Porto 2003). Thus, the central idea of this proposal was to interpret the *bios* of bioethics beyond the restrictive boundaries of biomedical ethics, and to incorporate the idea of *life* in its broadest sense.

The common background to this critical thinking which has been shared with other Latin American bioethics authors (Tealdi 2008; Vidal 2012), is the need to understand how the current process of globalization of human relationships is supported by an unfair world order (Cunha and Lorenzo 2014). Through reproduction of this process, power and income are becoming increasingly concentrated among groups that have historically been benefited by the center-periphery relationship of the world. This whole context allows perpetuation of unfair mechanisms for the sharing of benefits (in favor of a mostly white population in northern countries) and also allows the harm and ill effects that result from the process of globalization (to the detriment of a large portion of the world's population, and notably the black population of sub-Saharan Africa).

Regarding these critical approaches, although a significant proportion of the authors of bioethics production in the "north" do not take the ethical implications of maintaining the current global political and economic order into consideration, several "northern" scholars have structured their reflections based on different critical perspectives of globalization, such as Berlinguer (2000), Singer (2002), Benatar (2003), Lysaught (2009), Stone-Mediatore (2011), among others.

The present book itself expresses a certain critical identity regarding the traditional approaches of global ethics. It puts forward a more plural vision, directed towards overcoming global iniquities and injustices, through the diversity of authors and their own perspectives. In the opening page of the first chapter of this book ten Have presents an awkward question which forces us to seek adequate answers in this space so that we do not become passively complicit in injustices: is it possible to develop global ethics education, contributing to the formation of global citizens concerned with global health and justice, assuming global responsibility to criticize structures of violence and inequity?

Thus, by indicating a basis for constructing goals for global ethics, the reference point that we are taking is this more contextual and critical approach. This is especially prominent in the southern hemisphere but is also defended by northern authors. We emphasize that our interpretation is not merely geographical, regarding the place of origin of each author, but above all it is geopolitical. Before reaching this point, as mentioned previously, two current themes within the theoretical and practical context of a globalized world will be used as references: moral imperialism and coloniality. These themes serve as warnings that threats to adequate global ethics education need to be avoided. If these warnings are ignored, this system could possibly become another factual element serving to maintain the unilateral hegemonic interests of the current unjust global order, instead of being a democratic and socially inclusive instrument.

3.3 Global Ethics Education Opposing Moral Imperialism

To discuss the expressions "moral imperialism", which is used in bioethics more frequently than might be imagined, whether with a critical tone (Benatar 1998) or with a surprisingly positive one (Dawson and Garrard 2006), and "colonialism" (Chadwick and Schüklenk 2004), which is a term close to coloniality, the analysis needs to start from what is perceived as "power".

According to a broadened concept, there are three classes of power: economic, ideological and political. This last type is especially important for the present analysis. The classical sense of politics comes from the Greek *polis*, meaning everything that refers to a city and, thus, everything that is urban, civil, public and even sociable and social. The term has lost its original meaning and in modern times is used to indicate the activity or set of activities that in some way refers to the state.

The concept of politics is therefore understood as a form of human action that has a close relationship with power. The term power, on the other hand, can be interpreted as a relationship between two parties or subjects, in which one imposes its will over the other, thus determining the behavior that should be followed. Since domination over people cannot be considered as an end in itself, power as a type of relationship needs to be complemented with understanding of power as possession of the means that allow the desired results to be attained. What characterizes political power is the exclusivity of use of force (considering an immense variety of forms and intensities) in relation to the totality of human groups that act within a given social context (Bobbio et al. 2004). This exclusivity results from a process that takes place within organized societies, through monopolization of ownership and use of means that allow various forms of both implicit and explicit coercion: physical, economic, intellectual, psychological, religious, moral and so on.

The humanity of beings that are biologically recognized as humans is the consequence of a collective process that is consubstantiated through continuous production and reproduction of meanings attributed to social practices and practices that are vital for survival. Globalization of the economy, the planet's geography and worldwide relationships themselves has not decreased the distance between rich and poor. Rather, it has hardened contradictions and accentuated existing problems (Piketty 2013). In poor and developing countries (on the periphery), most of the population continues to struggle every day for the minimum conditions of survival with dignity. On the other hand, through indiscriminate use of certain concepts of applied ethics that were proposed and implemented by several rich countries (at the center), the process of depoliticization of moral conflicts has grown, regardless of how serious such issues may be. Over the last decades of the twentieth century, with contributions from certain theoretical lines of activity within international bioethics, ethics started to be used in many instances as a horizontal and aseptic tool, distant from the concrete reality of the daily lives of individuals and people, thus serving politically neutral readings and interpretations of conflicts.

In the relationship between politics and (bio)ethics, provision of humanity to beings that are biologically recognized as humans is consequent to a dialectic process that includes the collectivity. This is consubstantiated through continuous production and reproduction of meanings attributed to social practices, in addition to existence of the body itself in its materiality, as further discussed in the final topic of this chapter. Thus, incorporation of politics into the agenda of issues dealt with through bioethics relates to the assumption that politically adequate social action is an action with the capacity not only to transform social praxis by means of empowering citizens regarding their rights, but also to promote inclusion of people who have been historically marginalized and excluded from the process of global development (Garrafa and Manchola 2014).

To exemplify what is discussed above, it is worth recalling an important event that occurred at the end of the 1990s regarding international research on HIV/AIDS involving human beings in peripheral countries (also called poor or developing countries, etc.). This strongly rekindled the discussion on the ethical limits of biomedical science and its relationship with the globalized market. At that time, the public became aware that studies conducted especially in African countries were applying abusive methodologies that were not allowed in developed countries, where the ethical norms for control over experiments were more rigorous (Lurie and Wolfe 1997).

Back in the 1980s, Marcia Angell, the editor of the journal in which these studies were published, had already demonstrated concern about the possibility that international research in the field of medicine could produce new forms of imperialism, thus anticipating an ethical conflict that she would directly face herself some years later (Angell 1988). Use of ethical and methodological standards that were more rigid for studies performed in sponsoring countries and more flexible in the countries that were being studied (a practice known as double standards) was immediately rejected by some authors (Angell, 1997; Benatar, 1998; Garrafa and Prado 2001), but was vigorously defended by others (Varmus and Satcher 1997; Resnik 1998).

The events that occurred during that period left important lessons and also served as a warning to be heeded in setting goals for global ethics education. The aim would be to avoid transformation of such teachings into yet another instrument that would strengthen the creation and maintenance of various forms of moral imperialism and coloniality of power and knowledge (a subject that will be discussed below). Thus, a classification system that can be related to global ethics education has been proposed for different forms of moral imperialism, through analyzing the historical development of abuse from these cases of international research.

The first type, which was called "direct moral imperialism" (Garrafa and Lorenzo 2008), was identified from mobilization among developed countries, led by the United States, to "add flexibility" to international documents for regulation of ethics in clinical trials. Specifically, this related to the Declaration of Helsinki, which has been modified regarding some neuralgic topics, in accordance with the interests of large economic groups that sponsor international clinical trials (Hellmann et al. 2014).

In turn, "indirect moral imperialism" was more subtle and resulted from educational processes of teaching and learning sponsored by public and private institutions in developed countries with the objective of co-opting managers and researchers in weaker countries so that they would reproduce more "flexible" ethical standards (Garrafa and Lorenzo 2008). Expressing this more directly, this would allow loosening of rules and meet the interests of these sponsoring countries and their industries. In Latin America, for example, the legitimacy of training courses for researchers and members of both local and national committees on ethics in clinical research, conducted through educational programs promoted by the United States National Institutes of Health (NIH) and the Fogarty Foundation, has been questioned (Hellmann et al. 2015; Hellmann et al. 2016).

Once again, it is imperative to note that a critical approach towards the form of production of international ethical norms and educational programs on global ethics is not exclusive to authors in peripheral countries. This concern has also been expressed by several authors in the northern hemisphere. Lysaught (2009), for example, pointed out that the international normative regime in the field of research ethics sought to legitimize a form of "biopolitics". This would be maintained by means of strategies for surveillance and control of the wellbeing of individuals and populations, with the aim to produce "docile bodies" and even "docile states" that cooperate with the interests of countries that are considered developed.

However, despite repeated and exhaustive complaints regarding various forms of imperialist practices in the field of global ethics, recent evidence has shown that exploratory and abusive studies are still conducted in the peripheral regions of the world, including Latin America, where a considerably more robust research ethics system has been implemented, in comparison with other regions of the southern hemisphere (Homedes and Ugalde 2016).

Thus, educational programs on global ethics need to be committed to a critical and self-critical position in the light of wider hegemonic political and economic interests. These programs should be guided by new bases for defining overall goals that result in real benefits for the worldwide collectivity and not only for people who have historically benefitted from the rich but contradictory process of globalization.

3.4 Global Ethics Education in Opposition to Coloniality

Starting from the theoretical contributions of studies on coloniality, several authors in Latin America have more recently begun to pose the problem of the mechanisms of power, control and global domination. These involve various dimensions of human life and have a correlation with the forms of scientific production and international norms. Studies on coloniality question how the hegemonic epistemology of various fields tends to overrate the contributions of Euro-American thinking and tends to silence other manifestations of thought (going beyond language barriers), particularly the contributions from Africa, from the poorest areas of Asia and from Latin America (Rivas-Muñoz et al. 2015). Consideration of this social, cultural and geopolitical dimension is important in aiming towards better understanding of the various current global ethical conflicts, including problems relating to the growing numbers of international refugees, uncontrollable migrations, wars and ethnic persecutions that are seen around the world today. Thus, the basis for the present topic comes from the original research that introduced studies on coloniality into the sphere of bioethics and, specifically, intervention bioethics, which was developed by Nascimento and Garrafa (2011).

According to this study, it would be impossible to understand the effects, senses, dynamics and practices of power within modernity without understanding that it is based on a colonial logic. This logic indicates that there is a certain hierarchy between what is local and what is global, insofar as what is local is particular and needs to be developed to reach global hegemony. Therefore, modernity can be seen as a construction of a new world image (ordered, rational, predictable and in constant progress).

This image of the modern world is defined by studies on coloniality as a "Eurocentric mode of interpretation". Through this image, everything that is not modern is deemed uncivilized, and therefore is notably subordinate and marginalized. The local scale is, within this context, lesser, subordinate and marginal. Local matters are distant from the center and are seen as something that is retrogressive and as something that needs to be educated, improved and developed so it can in the future reach ideal or global characteristics (Mignolo 2003; Nascimento and Garrafa 2011).

This image supposes (and institutes) hierarchical ranking between those who are developed and those who are not, such that this hierarchy is thought of in terms of who is modern and who is not. Those that are marked as not being modern are almost naturally considered inferior, thus indicating that they need education, need civilization and need to be placed on the track towards progress (paved by those who are already modern and developed), even if this implies (as frequently occurs, thus serving the interests of the present discussion and review) establishment of a process of domination. This process can be either subtle and invisible or direct and palpable, depending on the situation. Part of the argument legitimizing colonization thus relates to the aim of civilizing undeveloped, uncivilized and unmodern/non-modernized societies, which persisted even after the end of colonialism and is maintained in the form of what is called coloniality (Nascimento and Garrafa 2011).

Aníbal Quijano, the main constructor of the concept of coloniality, stated that modernity as we know it only exists because it was installed as a form of exerting power to undermine other people, which generates identities through violent creation of otherness (Quijano 1992). Thus, coloniality is precisely the power regime that, based on an idea of development, imposes economic, political, moral and epistemological standards on other nations, not only to establish an expansion mechanism for developed nation states, but also particularly to create a global Eurocentric and Americanized identity.

Therefore, to think of knowledge and concepts (whether these are ethical, epistemological or political) from the point of view of another hemisphere or from another geopolitical orbit implies understanding that these concepts are immersed in colonial historicity. This has a fundamental role, since coloniality of knowledge is located within coloniality of power and it legitimates and controls this power (Quijano 2000; Segato 2013). This regime or state of coloniality of knowledge is what determines domination, submission and dependency, through controlling the content and forms of transmission and application of knowledge.

Thus, colonial logic regarding modernity not only is connected to the context of political coloniality (the manner of exerting power based on colonial differences, which hierarchizes experiences, knowledge, cultures and lives) but also supports a knowledge production regime that legitimates this process and makes it function. At the same time, this knowledge is connected to a certain image of life and management of life, as discussed by Michel Foucault in considering the idea of biopolitics (Nascimento and Garrafa 2011).

One of the characteristics of coloniality consists of thinking of real situations in terms of hierarchical structures. Not only should individuals and nations that are supposedly less developed be under supervision by more developed individuals and nations, but also the essence of development itself should in some way be related to this supervision. The different ways of life of central and peripheral countries, geopolitically represented as northern and southern, encompass not only an assumption of difference in their levels of development, but also a ranking of the value of lives. This variant of colonial difference is the foundation for "coloniality of life", which is the process of creating ontology of life that permits the thought that some lives are more important than others. This allows establishment of a hierarchy and justification for domination, exploitation and submission, under the pretext that this is the path for future development of less developed lives (Nascimento and Garrafa 2011).

It is in this context that coloniality of life starts to be seen in the study model presented here as a basis for hegemonic biopolitics and, for example, some theoretical currents of bioethics that are distant from reality and deliberately "neutral". In some places in the present globalized world, "a prudent and regulated conception of life" hinders stronger and even violent attacks. In others, which are "less developed", these attacks are perfectly legitimate and authorized. This is not merely a word game, because it involves concepts regarding what life is for various people and these people's relationships with political and social macro-levels. Arguments defending double standards in clinical research are common nowadays and naturalize social relationships in peripheral countries. These arguments give the idea that these lives may be more important for world development, since these are the ones that are forced most towards their limits such that they are used as research subjects not only for the rest of the world, but also for the benefit of countries where these individuals live, at least at the theoretical level of the argument. Therefore, colonialism is extinct in the current globalized world, but coloniality continues to exist, with application of the same unilateral logic as before, but now with more sophisticated methods than the formal presence of the metropole in the colonized country (Nascimento and Garrafa 2011).

Inclusion of the present topic in the context of this chapter is justified by the fact that, with so many advances in the technical-scientific field already reported at this point of the twenty-first century, the global world cannot overlook the importance of (or even the need for) for defining the collectivity as a central focus for the goals that will be established for global ethics education. At the same time, care needs to be taken such that this program does not become a tool used by historically colonizing countries applied with disproportionate strength to dominate the rest of the world.

Similar to what was said in relation to moral imperialism, understanding the problems ensuing from the viewpoint of coloniality makes it possible to define the goals that will be set for studying ethics at a global scale, exactly in contraposition to the reasons and mechanisms that contribute towards creation and maintenance of these undesirable and excluding situations.

3.5 Can Global Ethics Education Contribute Towards Overcoming Situations of Moral Imperialism and Coloniality?

Other current issues could have been used as guidelines for the present analysis, in the way that moral imperialism and coloniality were used. However, in addition to our strategic disposition towards developing a train of thought based on the point of view of the southern hemisphere, selection of these two subjects was based on their historical topicality and on the negative consequences from practices that have such an effect on the most needy populations of the contemporary world.

One question that necessarily emerges from this reflection is whether there is any possibility of establishing goals for global ethics education in the face of international situations on such a large scale, which involve so many sociopolitical variables and mega-interests. Such situations might be completely beyond the capacity and control of the most rigorous and exemplary educational programs that might one day be implemented.

This narrative should not, however, be interpreted as discourse of surrender. On the contrary, through seeking to locate the essence of the origin of these inequities, i.e. the global system of construction, control and distribution of material wealth that promotes higher or lower levels of human wellbeing and social inclusion or exclusion across the globe, this narrative renders another scenario that goes beyond simple criticism and enables not only congregation of reflections but also, especially, action.

At this point, we have reached the essence of the proposal of the present chapter. Within the current situation of the world, this chapter primarily affirms that it is highly important that entities, organisms, universities, groups and people who work either directly or indirectly with ethics education should understood that it is insufficient, and even harmful, to try to reach global goals defined aprioristically. However, it is possible to define a common basis founded on a commitment towards considering global ethics education as a concrete tool for identifying forces and power structures that determine global ethical conflicts, through considering local specificities. This is particularly so in relation to issues that determine the possibility of life with dignity and quality for all.

This means that global ethics education must, above all, be governed by an understanding that it is possible, through using the set of tools that global ethics can itself provide from a critical perspective, to act towards comprehension, revelation and deconstruction of the discursive, ideological and normative mechanisms that determine ethical conflicts at local and global scales and which are incompatible with the scientific knowledge and development that has been reached today. Thus, in seeking to surmount situations such as moral imperialism and coloniality, the existence of global mechanisms of power needs to be recognized. This search also necessarily implicates the locus where the drivers of this subjection and weakness (manifested by the incapacity of individuals and nations to separate themselves from submission) are installed. By defining these poles, Freire (2001) identified the contrast between captivity (or deprivation of the right of choice) and freedom, i.e. true exercise of autonomy. According to this author, social subjects, even those within academia or in global ethics education programs like the one espoused here, are eminently political agents, whose actions can either maintain or transform the *status quo*. Comprehension of this reflection reveals positions of power and allows the assumption that a position can be taken in the game of forces towards social inclusion based on awareness of the drivers of oppression, and towards organization of concrete actions in opposition to these.

On the other hand, based on the spaces and tools of global ethics, there is a need for resignification and for construction of new values and ethical reference points that are compatible with ethically adequate existence in this world for all groups, including people who have historically shouldered the burden of globalization. This means that the criticism of moral imperialism and processes of coloniality that were established in the present analysis do not hinder recognition of the need to seek a worldwide consensus in the field of teaching of global ethics. On the contrary, the existence of ethical conflicts caused by a global power structure that is ideologically supported on universal values (including, for example, the distorted values of freedom that support the discourse of the universality of the free market; or partial use of the discourse of human rights to justify invasion of countries that do not submit to the interests of central countries) needs to be faced globally.

This means that there is clear tension between the recognition that, on the one hand, these are global processes that within the sphere of power create and support conflicts regarding new forms of moral imperialism and coloniality, and the recognition that, on the other hand, it is possible and necessary to be able to count on global ethics education as a tool for comprehension of and coping with these and other problems. In this regard, global ethics education should be based on a universal axiological foundation that considers historically contextualized comprehension of the ethical phenomenon. Thus, the proposal is that the "human body" itself should be interpreted as a fundamental and universal ethical principle, considering that human beings share existence from a bodily perspective and that all other moral values and principles emanate, necessarily, from a body that can think, feel, move and relate to others (Garrafa and Porto 2003). As indicated previously, we began to collaboratively construct a theoretical approach called intervention bioethics some years ago. In this, the paradigm of corporeity is presented as one of the central elements grounding this approach, together with other categories of importance for global comprehension of ethics, including the distinction between emerging and persistent problems and the dynamics of the relationship between the center and the periphery of the globalized world (Garrafa 2000; Garrafa and Porto 2003; Nascimento and Garrafa 2011).

An understanding of the body as a universal principle, with interpretation of the pleasure and pain that emanate from it as somatic markers for actions that are ethically either ethically or undesirable, enables a universalizing approach to ethics without running into ethnocentric, abstract and decontextualized propositions. It should be noted that other authors within bioethics also based their studies on materialist perspectives regarding the human body in indicating global approaches to ethics. Potter, for example, defended health as a basic value for global ethics insofar as, in his view, no "nation, tribe, or religious community has ever regarded malnutrition, parasitism, or disease of any kind as a desirable goal" (Potter 1988, p. 155).

Therefore, through materializing the dynamics of globalization in the body, a solid moral foundation promoting the basis for constructing goals for global ethics education can be encountered. This foundation comes from comprehension that we live in a global context in which experiences of pleasure and pain (and the ability to live healthily and avoid diseases) do not depend solely on individual decisions, or even decisions of particular governments. These experiences depend on complex political, economic, social and cultural relationships that have direct and indirect repercussions on systematic and global phenomena such as poverty, unemployment, different forms of exclusion, climate change, unequal use of the world's limited natural resources, etc. Moreover, support for globally shared ethical responsibilities originates from the recognition that Earth is a geologically closed and interdependent system that is invariably used as a shared asset by all its current or future inhabitants, including both human and non-human beings (Cunha 2014). Thus, for example, it is not an overstatement to report that in the Brazilian Amazon region, there are today more than ten different indigenous groups that have never been contacted by what is called "civilization". Moreover, although these groups have no relationship with the global economy, the environmental effects caused by abuses already affect them.

Taking the paradigm of corporeity as a reference, among other possibilities, it is believed that there is a concrete possibility of creating an adequate space for development of feasible goals for global ethics education. Thus, this scenario will be properly shaped so as to contribute towards overcoming the major ethical conflicts that directly affect the daily lives of most of the Earth's inhabitants. At the same time, the existence of this scenario inhibits reproduction of incursions within the sphere of global ethics education that favor (re)appearance of undesirable forms of domination, such as moral imperialism and coloniality, which were taken as references for the present chapter.

3.6 Final Remarks

To avoid new processes of exploitation and domination through actions carried out under the insignia of global ethics, a critical commitment towards surmounting the international economic system that perpetuates iniquity between developed countries and the rest of the world needs to be adopted, with a critical reading of educational actions regarding global ethics.

Thus, it is important to seek a form of global ethics education that is governed by critical categories that can contribute towards establishment of a global moral conscience that supports a concrete commitment towards searching for solutions for ethical conflicts that affect humanity, particularly those relating to processes of exploitation and domination.

Among the possible critical categories for a program of global ethics education, the present chapter proposed consideration of moral imperialism and coloniality as notions that, in addition to allowing contextualized reading of globally determined ethical conflicts and their consequences, will allow reference marks that inhibit use of global ethics education as a tool for maintenance and expansion of global political and economic processes that would increase social inequities, cultural conflicts and environmental degradation.

Finally, instead of proposing goals that were set aprioristically from abstract ethical principles, this chapter indicated the paradigm of corporeity as a universal marker for what should either be morally aspired to (pleasure and health) or avoided (pain and disease) in educational programs on global ethics, thereby providing a minimum axiological basis upon which goals can be established. We therefore hope that critical consideration of the power relationships that form geopolitical dynamics and a commitment of respect towards corporeity answer the two great challenges that were addressed in this chapter: recognition, within the field of global ethics education, that it is necessary to establish a global consensus that respects local particularities; and a commitment to make global ethics a tool that will effectively contribute towards overcoming the main ethical conflicts that affect mankind.

References

Angell, M. 1988. Ethical imperialism? Ethics in international collaborative clinical research. *The New England Journal of Medicine* 319: 1081–1083.

———. 1997. The ethics of clinical research in the third world. *The New England Journal of Medicine* 337 (12): 847–849.

Arnold, D.G., ed. 2014. *The Ethics of Global Climate Change*. Cambridge: Cambridge University Press.

Beauchamp, T.L., and J.F. Childress. 2013. *Principles of Biomedical Ethics*. 7th ed. New York: Oxford University Press.

Benatar, S. 1998. Imperialism, research ethics and global health. *Journal of Medical Ethics* 24: 221–222.

Benatar, S.R. 2003. Improving global health: The need to think "outside the box"! *Monash Bioethics Review* 22 (2): 69–72.
Benatar, S. 2011. In *Global Health and Global Health Ethics*, ed. G. Brock. Cambridge: Cambridge University Press.
Berlinguer, G. 2000. *Bioetica quotidiana*. Firenze: Giunti.
Bobbio, N., N. Matteucci, and G. Pasquino. 2004. *Dicionário de Política*. Brasília: Editora Universidade de Brasília.
Chadwick, R., and U. Schüklenk. 2004. Bioethical colonialism? [editorial]. *Bioethics* 18: iii–iiv.
Cunha, T.R. 2014. *Critical bioethics, Global Health and the development agenda*. Thesis, Unesco Cathedra in Bioethics, University of Brasília, Brazil.
Cunha, T., and V. Garrafa. 2016. Vulnerability: A key principle for global bioethics? *Cambridge Quarterly of Healthcare Ethics* 25 (2): 197–208. https://doi.org/10.1017/S096318011500050X.
Cunha, T., and C. Lorenzo. 2014. Global bioethics from the perspective of critical bioethics. *Revista Bioética* 22 (1): 116–125. https://doi.org/10.1590/S1983-80422014000100013.
Dawson, A., and E. Garrard. 2006. In defense of moral imperialism: Four equal and universal prima facie principles. *Journal of Medical Ethics* 32 (4): 200.
Engelhardt, H.T., ed. 2006. *Global Bioethics: The Collapse of Consensus*. Salem: M&M Scrivener Press.
Fortes, P.A.C. 1998. *Ética e saúde*. São Paulo: EPU.
Freire, P. 2001. *Pedagogia da Autonomia*. Paz e Terra: São Paulo.
Garrafa, V. 1995. *Dimensão da ética em saúde pública*. São Paulo: Faculdade de Saúde Pública USP/Kellogg Foundation.
———. 2000. Bioética fuerte: una perspectiva periférica a las teorías bioéticas tradicionales. In *3er Congreso de la Federación Latino-Americana y del Caribe de Instituciones de Bioética – FELAIBE*. Panamá.
Garrafa, V., and C. Lorenzo. 2008. Moral imperialism and multi-centric clinical trials in peripheral countries. *Cadernos Saude Publica* 24 (10): 2219–2226. https://doi.org/10.1590/s0102-311x2008001000003.
Garrafa, V., and C. Manchola. 2014. Bioethics: A tool for peacebuilding. *Revista Colombiana de Bioética* 9 (2): 95–106.
Garrafa, V., and D. Porto. 2003. Intervention bioethics: A proposal for peripheral countries in a context of power and injustice. *Bioethics* 17 (5–6): 399–416. https://doi.org/10.1111/1467-8519.00356.
Garrafa, V., and M.M. Prado. 2001. Changes in the declaration of Helsinki: Economic fundamentalism, ethical imperialism and social control. *Cad. Saúde Pública* 17 (6): 1489–1496.
Garrafa, V., D. Diniz, and D.B. Guilhem. 1999. Bioethical language and its dialects and idiolects. *Cadernos Saúde Pública* 15 (Suppl. 1): 35–42. https://doi.org/10.1590/S0102-311X1999000500005.
Heinze, E.A. 2016. *Global Violence: Ethical and Political Issues*. London: Routledge.
Hellmann, F., M. Verdi, S. Bruno, and S. Caponi. 2014. 50th anniversary of the Declaration of Helsinki: the double standard was introduced. *Arch Med Res* 45 (7): 600–601.
Hellmann, F., V. Garrafa, B.R. Schlemper Junior, and S. Bittencourt. 2015. The Fogarty Training Program in low- and middle-income countries: international research ethics education or moral imperialism? *Archives of Medical Research* 46 (6): 515–516. https://doi.org/10.1016/j.arcmed.2015.08.003.
Hellmann, F., V. Garrafa, B.R. Schlemper Junior, and S. Bittencourt. 2016. Moral Imperialism in NIH Fogarty Training Program and the continuing unethical trials in poor countries: a rejoinder. *Archives of Medical Research* 47 (1): 67–68.
Holm, S. 1995. Not just autonomy: the principles of American biomedical ethics. *Journal of Medical Ethics* 21: 332–338.
Homedes, N., and A. Ugalde. 2016. Health and ethical consequences of outsourcing pivotal clinical trials to Latin America: A cross-sectional, descriptive study. *PLoS ONE* 11 (6): 1–17. https://doi.org/10.1371/journal.pone.0157756.
Jonsen, A.R. 2003. *The Birth of Bioethics*. New York: Oxford University Press.

Lurie, P., and S. Wolfe. 1997. Unethical trials of interventions to reduce perinatal transmission of the human immunodeficiency virus in developing countries. *The New England Journal of Medicine* 337 (12): 853–856.
Lysaught, M.T. 2009. Docile bodies: Transnational research ethics as biopolitics. *Journal of Medicine and Philosophy* 34 (4): 384–408. https://doi.org/10.1093/jmp/jhp026.
Macklin, R. 1998. Ethical relativism in a multicultural society. *Kennedy Institute of Ethics Journal* 8 (1): 1–22.
Mainetti, J.A. 2010. The Discourses of Bioethics in Latin America. In *Ibero-American bioethics – History and perspectives*, ed. L. Pessini, C. de Paul de Barchifontaine, and F. Stepke, 21–28. London/New York: Springer.
Mignolo, W. 2003. *Histórias locais/Projeto globais: colonialidade, saberes subalternos e pensamento liminar*. Belo Horizonte: EdUFMG.
Moellendorf, D., and H. Widdows. 2014. Global ethics: A short reflection on then and now. *Journal of Global Ethics* 10 (3): 319–325.
Nascimento, W.F., and V. Garrafa. 2011. For a not colonized life: Dialogue between intervention bioethics and coloniality. *Saúde Soc. São Paulo* 20 (2): 287–299. https://doi.org/10.1590/S0104-12902011000200003.
Piketty, T. 2013. *Le capital aux XXI*[e.] *siécle*. Paris: Éditions de Seuil.
Pogge, T. 2008. *World Poverty and Human Rights*. 2nd ed. Cambrigde: Polity Press.
Potter, V.R. 1971. *Bioethics. Bridge to the future*. Englewood Cliffs: Prentice Hall.
———. 1988. *Global Bioethics: Building on the Leopold Legacy*. East Lansing: Michigan State University Press.
Potter, V.R., and L. Potter. 1995. Global bioethics: converting sustainable development to global survival. *Medicine and Global Survival: M & GS* 2 (3): 185–191.
Quijano, A. 1992. Colonialidad y modernidad/racionalidad. In *Los conquistados*, Bonillo, H. (Comp.), 437–449. Bogotá: Tercer Mundo/Flacso.
———. 2000. Colonialidad del poder y clasificación social. *Journal of World-Systems Research*. Santa Cruz XI (2): 342–386. Special issue.
Rendtorff, J.D., and P. Kemp. 2000. *Basic Ethical Principles in European Bioethics and Biolaw: Autonomy, Dignity, Integrity and Vulnerability*. Barcelona: Institut Borja de Bioética.
Resnik, D. 1998. The ethics of HIV research in developing nations. *Bioethics* 12: 286–306.
Rivas-Muñoz, F., V. Garrafa, S.F. Feitosa, and W.F. Nascimento. 2015. Bioethics of intervention, inter-culturality and non-coloniality. *Saúde Soc. São Paulo* 24 (supl.1): 137–147.
Schramm, F. 1996. *Bioética – a terceira margem da Saúde*. Brasília: Editora UnB.
Segato, R.L. 2013. *La crítica de la colonialidad en ocho ensayos y una antropología por demanda*. Buenos Aires: Prometeo Libros.
Singer, P. 2002. *One World Now – The Ethics of Globalization*. New Haven: Yale University Press.
Stone-Mediatore, S. 2011. A not-so-global ethics contradictions in U.S. global ethics education. *Philosophy in the Contemporary World* 18 (1): 43–57.
Tealdi, J., ed. 2008. *Diccionario Latinoamericano de Bioética*. Bogotá: Unibiblos/Unesco.
ten Have, H., ed. 2015. *Bioethics Education in a Global Perspective*. 1st ed. Dordrecht: Springer.
———., ed. 2016. *Encyclopedia of Global Bioethics*. Cham: Springer.
Unesco. 2005. *Declaração Universal sobre Bioética e Direitos Humanos*. Retrieved from http://bvsms.saude.gov.br/bvs/publicacoes/declaracao_univ_bioetica_dir_hum.pdf.
Varmus, H., and D. Satcher. 1997. Ethical complexities conducting research in developing countries. *New England Journal of Medicine* 337: 1000–1005.
Vidal, S., ed. 2012. *La Educación en Bioética en América Latina y el Caribe: experiencias realizadas y desafíos futuros*. Montevideo: UNESCO. Retrieved from http://unesdoc.unesco.org/images/0022/002255/225533s.pdf.

Chapter 4
Priorities in the Teaching of Ethics in a Globalized World

Renzo Pegoraro

Abstract Globalization has created the favorable conditions for social, economic and cultural integration in today's world; but also problems and challenges involving ethics and bioethics. In this context, beside the many benefits of global range, several challenges have emerged, requiring a specific reflection. These challenges involve general disciplines such as anthropology, medicine/healthcare, and ecology as well as specific aspects: in a broad range, those linked to social justice fulfillment, guaranteeing safety, promoting interreligious dialogue, building peace; in a narrow range, those related to professional moral conduct, healthcare delivery, access to healthcare and equitable distribution of healthcare resources, and provision of treatment.

Therefore, new priorities arise also in the teaching of ethics and bioethics in a globalized world.

"Global conscience": first of all; we have the necessity of providing moral knowledge, skills and attitudes, developing consciences to have a new ethical awareness, enabling first to recognize and then to face the challenge posed by the globalized world.

"Universal values": second; we must spread universal values, particularly anthropological values such as human dignity (i.e. referring to those documents that have a certain consensus, such as the *Universal Declaration on Bioethics and Human Rights*, UNESCO, Universal declaration on bioethics and human rights. Retrieved from: http://portal.unesco.org, 2005).

"Common responsibility": third; rethink and propose again responsibility in seeking a sustainable and integral development. Common (or global) responsibility must take into account both promotion of human beings' dignity and preservation of creation, showing how all these aspects are absolutely interconnected (integral ecology).

R. Pegoraro (✉)
Pontifical Academy for Life, Vatican City, Rome, Italy

Fondazione Lanza, Padua, Italy
e-mail: renzo.pegoraro@fondazionelanza.it

H. ten Have (ed.), *Global Education in Bioethics*, Advancing Global Bioethics 10,
https://doi.org/10.1007/978-3-319-78984-2_4

4.1 Introduction

The phenomenon of globalization is continually expanding and with growing complexity, marked by positive aspects and by problematic aspects. Faced with this recent and rapid phenomenon, even bioethical reflection is called to reconsider itself, reviewing its approach, which is often too "Western-oriented," and its concentration only in the medical-clinical sphere (Ten Have and Gordijn 2014).

The development of a global ethics is always most important to accompany and intervene in the process of globalization. This also implies a rethinking of the content and of the methods of teaching ethics and bioethics in this globalized world.

A broad, international look that embraces all the determinants of health can help to define the priorities of this teaching for a truly human, moral, and cultural education. There should be an opening up and change of mentality to offer greater attention to the issues of justice and solidarity in the human community and in the relationship with the environment. This extends also to improve actions for a change of economic, social, and health structures for the true good of everyone. It is necessary to understand the challenges and try to propose an education in bioethics that develops the new "big ideas" and concrete actions.

4.2 Reality and Challenges in a Globalized World

Globalization has created favorable conditions for social, economic and cultural integration in today's world. The development of information and communication technologies that facilitate connecting people across countries has certainly made possible an overall interconnection among nations and regions and has made easier the movement of people, goods and services across borders.

Economic interaction has rapid and relevant consequences among all countries in the world and in their social conditions, markets, welfare and health care policies. Attempting to recognize some of the general trends in the current globalized world, we can draw attention to the need for energy and raw materials, the need to solve the problem of hunger, the incessant fight against diseases and the protection of the health of humans, animals, and plants (included in the perspective of global health and the promotion of healthy lifestyles, not only prevention of diseases), the aspiration within the population to achieve a new quality of life, and the search for new technological platforms of innovative development in the conditions of global competition (Nezhmetdinova 2013).

In this context, alongside the many benefits of global range, several challenges – exacerbated by the diversity of perspectives in a society characterized by a pluralism of moral visions – arise, requiring more specific reflection, but the speed of social and cultural change is so rapid that humankind does not have enough time to comprehend and learn (Nezhmetdinova 2013). These challenges involve general fields like anthropology, medicine/healthcare, and ecology as well as more specific aspects: in a "broad range," those linked to attaining social justice, guaranteeing safety, promoting interreligious dialogue, and peace-building, and in a "narrow range" those related to professional moral conduct, healthcare delivery, access of healthcare and equitable distribution of healthcare resources, and provision of treatment.

The most relevant problem appears to be the dominance of a neoliberal ideology that privileges the individual over society and justifies a so-called "market fundamentalism" that heavily affects the justice and health of much of the human population (Ten Have 2016; Gracia 2014). Regarding the sphere of public health, while globalization has facilitated improvements in health care, it has also created new hazards and avenues for exploitation. Globalization has led to a sharing of both risks and responsibilities in public health; e.g. the ease with which infectious diseases can spread within the global community (given the speed of modern travel and trade), and the different ways to combat them in developed and developing countries. Many public health problems exceed national barriers, involving environmental issues such as pollution and water scarcity and socio-economical factors such as poverty, malnutrition, and refugees. All "global" problems affect regional and local levels; in turn, "local" problems can have a worldwide impact.

4.3 Setting New Priorities in the Teaching of Ethics

As a result of globalization, all these issues can no longer be understood solely within the political boundaries that define traditional notions of individuals and communities. Solutions demand a global conception of rights and obligations, which in turn requires a new approach and a reevaluation of existing ethical and legal frameworks.

In the prospect of ethics education in this global view, it is important to recognize a more integrated approach by recovering the interdisciplinary setting typical of bioethics, with greater attention paid to economic, environmental, and social-cultural aspects. It means paying greater attention to social ethics and bio-political ethics.

It is important in the effort of teaching bioethics about global health "to shift the emphasis from sensational cases to everyday matters, from autonomy to justice and from access to healthcare to the social determinants of health" (Dwyer 2003, p. 432). Therefore, the problem of inequalities, the duty to assist, the strategies for the prevention of diseases, malnutrition, become very relevant and it is urgent to recognize the ethical implication of all these problems, and to teach ethics for these "global-local" problems.

The perspective is to develop an educational process that takes into account the two levels of a global bioethics: "On an abstract level, there is a set of minimum standards, on which different traditions and culture agree. On a more contextualized level, there are efforts to articulate more specific bioethics standards in the context of specific religious and cultural traditions. In addition, local specifics of cultures and traditions are important for the interpretations and application of universal standards. The two levels are interacting along bottom up and bottom down lines of communication. Global platforms and local contexts mutually help each other to construct and produce global bioethics. Thus, global bioethics is the result of continuous and multilateral articulation, deliberation, and production" (Ten Have and Gordijn 2014, pp. 16–17; Kymlicka, 2007).

Furthermore, it is important to define better the objectives, methods, content, and evaluation of a proposal of bioethics education, with all the problems related to teaching in this field (Ten Have and Gordijn 2014, pp. 447–466; Hellsten, 2015). A significant and relevant proposal is that produced by UNESCO with the development of a core course in bioethics, designed around the bioethical principles of the *Universal Declaration on Bioethics and Human Rights*, adopted by UNESCO in 2005, in the perspective of human rights and global bioethics (UNESCO 2008, 2005). An interesting contribution to a bioethics education in a global perspective is offered by an international publication edited by Henk ten Have, presenting global perspective and comparative analysis of experiences in teaching bioethics in a globalized world. However, it is possible to try to define new priorities that arise in the teaching of ethics and bioethics in a globalized world (Ten Have 2015). They can be summarized as follows:

Creating a "global conscience,"
Spreading "universal values,"
Sharing "common responsibility."

(a) *Creating a "Global Conscience"*

First of all, we need to provide moral knowledge, skills and attitudes to develop consciences capable of new ethical awareness and able to recognize the new challenges posed by the globalized world and to face them properly. For example, analyzing possible positive and negative impacts of global problems on local communities; recognizing that the gap between developed and developing countries raises important questions about the meaning of rights and obligations in an international context; and solving conflicts between public interests and individual rights.

When it comes to bioethics, much discourse until now has focused on high-tech issues (such as stem cell research, and nanotechnologies) or on traditional issues (e.g. informed consent and individual autonomy).

Bioethics should now instead focus on issues that affect a greater majority of people in the world, contributing to create a "global conscience" for "global problems." Schüklenk and Bello propose, to that end, a range of topics that should receive greater attention by professional bioethicists:

- the "10/90 gap" in health research, that is, the finding by the Commission on Health Research for Development in 1990 that less than 10% of worldwide resources devoted to health research were put towards health in developing countries, where over 90% of all preventable deaths worldwide occurred (Global Forum for Health Research, 2004),

- the transnational organ trade,
- access to essential medicines,
- health-based immigration restrictions,
- international research ethics and the flow of health information (Schüklenk and Bello, 2006).

This global conscience should characterize individuals and their communities, involve healthcare professionals and health policy-makers, and recognize the importance of all determinants of health and the responsibility of global action.

In the teaching and education experience, therefore, individuals must be helped to become aware of belonging to the "common home" and to feel directly engaged. At the personal as well as community levels (family, State, profession…), this "global conscience", which values dialogue and active participation, can be founded on relatedness (among people and with nature). It is a "hermeneutical cycle" that develops starting from knowing and discerning the local realities for a global vision and action, which sustains local reconversion and initiative. This circularity must be highlighted and accomplished, being aware of the values that inspire our discernment and our responsibilities.

In the perspective of "globalization", there is awareness of including not only all actual human beings, but also future generations and all other leaving organism and also environment. Teaching and education in ethics should move in this "inclusive" direction. In the field of global bioethics it becomes relevant to integrate the basic values and principles of bioethics, medical and nursing professionalism, the different types of ethics committee, principles inspiring health care institutions (UNESCO 2008). Therefore, all people involved in the ethical issues, could develop a "global conscience" and participate in it, through the personal and professional conscience, in this open and integrated perspective.

(b) *Spreading "Universal Values"*

Second, we must spread universal values, particularly anthropological and moral values such as human dignity, justice and solidarity. For this, it is important and useful to refer to those documents that have a certain international consensus (such as the *Universal Declaration on Bioethics and Human Rights*, UNESCO 2005).

Therefore, sensitivity and commitment must be encouraged to ensure that universal values are recognized in every country (especially in those countries where human life is threatened and systematic violation of fundamental rights is permitted).

Then, it is necessary to develop the capacity to render universal values into the context of a local culture, "translating" them into cultural categories already present in regional areas.

Everything is to be done in the prospect of dialogue and integration between cultural pluralism and minimal bioethical standard. "Establishing a global community of shared values should be possible, by considering them as the outcome of an ongoing process of interpretation, negotiation and dialogue. These values do not simply override the diversity of cultures and traditions, but - at least in part – they can be harmonized with them through the interpretation. Interpretations will never eliminate ethical disagreement, of course, but they can reduce the gap between the different starting points that derive from cultural traditions. In other words, the setting of common principles at the international level can proceed along with the recognition of deep cultural diversities" (Macioce 2016, p. 401).

In this perspective, religions can play a key role in spreading universal values, becoming a sounding board for the *Universal Declaration of Human Rights* of the U.N. and for the *Universal Declaration on Bioethics and Human Rights* of UNESCO. Religions indeed can provide ethical reference, meaningful values, and basic human needs that purchase permanent, irreversible, and unconditional effectiveness and can often reach a codification, even a legal one (as human rights or fundamental rights) (Küng 1987). There are problems concerning the dialogue and consensus among the religions and between religions and bioethics, but it is possible to find certain common elements. We can also consider the Golden Rule (treat your peer as you wish (s)he treated you) as a "cross-cultural principle, a contribution of the religion to the globalization of bioethics. It promotes equality, respect, non-violence and is largely applicable" (Sinaci 2016, p. 304).

This possibility is provided thanks to a factor common to all religious beliefs: "the inner imperative of the moral conscience, which enjoins us to respect, protect and promote human life, from the womb to the deathbed, for individuals and peoples, but especially for the weak, the destitute, the derelict: the imperative to overcome selfishness, greed and the spirit of vengeance" (Pope John Paul II 1986, n. 4).

In the educational experience in ethics, special and urgent attention must be devoted to the principle of solidarity. Such a principle implies mutual recognition of the other as an individual, within society; and in the international community among countries and peoples (Ricoeur 2004).

Such recognition sustains an awareness of and a commitment to the ties existing among generations, the different social components, Countries, the different national and supranational institutions, the different religions.

The attention for solidarity steers towards forms of collaboration and synergy to reach concrete results in the face of large and highly complex problems. The ethics of solidarity contrasts with an ideological vision according to which only competition allows for social and economic development, based on a free market and a neoliberal vision.

Instead, there is an urgent need to display and uphold a vision that favours collaboration as the most viable way for social and economic development, integrating and maximizing resources and capabilities, sharing objectives and means, in a spirit of alliance, given the great and demanding challenges that stand before us. It is thus possible to address the serious inequalities that afflict humanity: the poverty of many nations, the lack of food and water, the causes of pollution, the difficulty in ensuring equal access to adequate healthcare. There is a great ethical responsibility that must be taught and formed, to cultivate solidarity and justice, so that all human beings may watch over the "common home" and look after the entire human family.

(c) *Sharing "Common Responsibility"*

The third issue is to rethink and re-propose responsibility in seeking a sustainable and integral development. Common (or global) responsibility must take into account both the promotion of human dignity and the preservation of creation.

Educational activities should teach a sense of "global community", with all people involved in the same responsibility, in the effort of a positive collaboration to develop instruments of solidarity and promotion of human rights. Respect for human life, justice, peace and respect for the environment are absolutely interconnected themes, and for this reason the common responsibility should be based and developed in the shape of an "integral ecology" (Pope Francis 2015).

In the effort to prevent some elements of an "integral ecology" it is possible to recall as follows. "Ecology studies the relationship between living organisms and the environment in which they develop. This necessarily entails reflection and debate about the conditions required for the life and survival of society, and the honesty needed to question certain models of development, production and consumption. It cannot be emphasized enough how everything is interconnected" (Pope Francis 2015, n. 138).

"We are faced not with two separate crises, one environmental and the other social, but rather with one complex crisis which is both social and environmental. Strategies for a solution demand an integrated approach to combating poverty, restoring dignity to the excluded, and at the same time protecting nature". (Pope Francis 2015, n. 139). Social ecology is necessarily institutional, and gradually extends to the whole of society, from the primary social group of the family to the wider local, national and international communities.

This prospect of integral ecology invites one to overcome an individualistic vision of the person and the pretension of solving global issues with economic rules or only through technique. It is not about exporting/importing models but enhancing the participation and assumption of responsibility of everyone and their corresponding communities, across their own cultural traditions. "A consumerist vision of human beings, encouraged by the mechanisms of today's globalized economy, has a leveling effect on cultures, diminishing the immense variety which is the heritage of all humanity. Attempts to resolve all problems through uniform regulations or technical interventions can lead to overlooking the complexities of local problems which demand the active participation of all members of the community. New processes taking shape cannot always fit into frameworks imported from outside; they need to be based in the local culture itself. (…) Nor can the notion of the quality of life be imposed from without, for quality of life must be understood within the world of symbols and customs proper to each human group" (Pope Francis 2015, n. 144).

Therefore an integral ecology is inseparable from the notion of the common good, a central and unifying principle of social ethics. The common good is "the sum of those conditions of social life which allow social groups and their individual members relatively thorough and ready access to their own fulfillment." (Second Vatican Council 1965, n. 26). The principle of the common good supports the respect for the human person as such, endowed with basic and inalienable rights ordered to his or her integral development. It has also to do with the overall welfare of society and the development of a variety of intermediate groups, applying the principle of subsidiarity. Finally, the common good calls for social peace, the stability and security provided by a certain order.

"In the present condition of global society, where injustices abound and growing numbers of people are deprived of basic human rights and considered expendable, the principle of the common good immediately becomes, logically and inevitably, a summons to solidarity and a preferential option for the poorest of our brothers and sisters" (Pope Francis 2015, n. 158).

As some have recalled, one could therefore speak of a logic of "convocation", that is a task shared by the entire human family, bringing to bear different energies and sensitivities (Morandini, 2017).

In a world strongly marked by the market logic and the power of technology, it is essential that there be ample participation on everyone's part with assessment and decision-making criteria based on the intrinsic dignity of every human person, on the quest for justice, on the idea of governance, limit, and respect.

Such a joint responsibility is based on a common hope and constantly fuels hope. In fact, the temptation to surrender and renounce, or to resort to violence in order to change things, is contrasted by the need to educate to steadfastness and strong commitment, to insist and not lose heart, to have hope even when much time is required and the obstacles that stand before us are extremely difficult.

An "ethics of hope", which is mutually driven by a "hope of ethics", based on confidence in the single human person and the whole human family.

4.4 A Critical and Prophetic Voice for a Fairer World

We must acknowledge the risk that bioethics might become merely one more academic subject and lose its characteristic capacity for creating "bridges" between different fields of knowledge and action and between the academic-scientific world and society. Indeed, bioethics must retain its "prophetic," critical and stimulating voice, which it can direct towards medicine and society itself.

Related to this last challenge, finally, is one of bioethics' most enduring tasks: the need to attend to the weakest and most vulnerable and to strive for justice and for rightful healthcare systems that can appropriately meet the needs and interests of each generation, including the most impoverished (Turoldo, 2007).

It is important to transmit a "passion" for life and for justice, with the force to create a real "bridge to the future," as expressed by V.R. Potter (Potter 1971; Potter 1988). Only in this way, with this global approach, can one also consider the specific problems in the clinical setting of a hospital, the problems of clinical ethics, the use of technology, the difficulty of communication between physician and patient, and the development and sharing of medical research (Pegoraro et al. 2007). It is one of the priorities in the teaching of bioethics to educate a critical attitude, a voice of reporting and encouraging, to find the good for everyone.

Therefore, a bioethics that is not "levelled out" or "adapted" to the System and to the existing power and market logics, with their negative consequences on the most vulnerable people. It is important to educate to an ethics and a bioethics that are capable of denouncing not only the behaviours of single individuals persons but the actual "structures of injustice" that cause damage and offense to people, that produce disease and prevent the necessary cures.

There is a growing need to dedicate more attention to social justice and to national and international healthcare organizations, and ethics should support this need.

4.5 Conclusion

Emerging social phenomena, the severe environmental crisis, globalization of information, and economic dynamics exhort us to elaborate on conceptual instruments and guidelines that can create an ethics capable of tackling various complex phenomena.

Teaching ethics in a globalized world is a fundamental and urgent challenge for institutions, healthcare professionals, teachers, students, and religions. It is a quest to develop a "real ethics for life," for everybody. "Combining a theoretical commitment with practical engagement implies the following: on the intellectual front, global bioethics should focus on further philosophical solidification and analysis of the foundations and basic concepts of human rights theory. One the practical front, it should promote ethics education, capacity building and protection" (Ten Have and Gordijn, 2014, p. 843).

In conclusion, teaching ethics in a globalized world requires a different approach and paradigm shift. It is necessary to have a broad vision of the problems at hand and to recognize how they are deeply interconnected. This requires accepting the complexity of phenomena and situations, facing the current challenges with discernment, prudence, but also courage. This requires developing a "global conscience" in order to feel that one's own responsibility is united to the responsibility of all the subjects and institutions concerned. Likewise, necessary is formation and an adequate interdisciplinary methodology that is capable of promoting dialogue among the new forms of knowledge, the various levels (local and international), and the different responsibilities for an ethics that pays attention to values and to the regulations to be drafted.

A new "casuistry" that is attentive to specific problems, but within such a global horizon of awareness and responsibility, will be ever more urgent in the hermeneutical cycle that helps to understand and act at the local and global level.

All this requires formation to dialogue and collaboration, authentic training in theoretical reflection, but also ways of analysing and putting forward ethical guidelines that may steer action. An ethics, therefore, that can comprehend the positions and contributions of the various sciences and the dimensions that problems present in order to reassert the universal values and human rights that guarantee the common good, while respecting and promoting every individual, especially the most fragile and in distress.

"New lifestyles" can thus be proposed and forged, which can be adopted and experienced by people and communities, that is personal and social "virtues" that can help to make this world of ours better.

"The greatest power of the biomedical, social, and environmental science is their capacity to shape the way we as human beings understand ourselves and the world in which we live. At one level – the most apparent – they give us new choice and thus new moral dilemmas. At another level, however they force us to confront established views of our human nature, and thus to ask what we should be seeking: What kind of people do you want to be? … The boundaries of bioethics cannot readily be constrained. The expanding boundaries force us to take up larger and deeper problems, much as a small stone tossed into the water creates larger and larger ripples." (Callahan 2004, p. 285).

References

Callahan, D. 2004. Bioethics. In *Encyclopedia of Bioethics*, ed. S. Post, 3rd ed., 278–287. New York: Macmillan Reference.

Dwyer, G. 2003. Teaching global bioethics. *Bioethics* 17 (5–6): 432–446.

Global Forum for Health Research. 2004. Retrieved from: www.globalforumhealth.org.

Gracia, D. 2014. History of global bioethics. In ten Have H. & Gordijn B. (Eds.), Handbook of Global Bioethics, Dordrecht: Springer, pp. 19–34.

Hellsten, S.K. 2015. The role of philosophy in global bioethics. *Cambridge Quarterly of Healthcare Ethics* 24 (2): 185–194.

Küng, H. 1987. *Theologie im Aufbruch: Eine ökumenische Grundlegung*. München/Zürich: Piper.

Kymlicka, W. 2007. The globalization of ethics. In *The Globalization of Ethics. Religions and Secular Perspectives*, ed. W.M. Sullivan and W. Kymlicka, 1–16. New York: Cambridge University Press.

Macioce, F. 2016. Balancing cultural pluralism and universal bioethical standards: A multiple strategy. *Medicine, Health Care and Philosophy* 19 (3): 393–402.

Morandini, S. 2017. Ecologia integrale tra scienza e teologia. In *Laudato sì*, ed. H.M. Yanez, 49–68. Roma: Gregorian and Biblical Press.

Nezhmetdinova, F. 2013. Global challenges and globalization of bioethics. *Croatian Medical Journal* 54 (1): 83–85.

Pegoraro, R., G. Putoto, and E. Wray. 2007. *Hospital Based Bioethics. A European Perspective*. Padova: Piccin.

Pope Francis. 2015. *Encyclical letter "Laudato sì'"*. Retrieved from: www.vatican.va.

Pope John Paul II. 1986. *Address to Representatives of the Christian Churches and Ecclesial Communities gathered in Assisi for the World day of prayer*. Retrieved from: www.vatican.va.

Potter, V.R. 1971. *Bioethics: Bridge to the Future*. Englewood: Prentice-Hall.

———. 1988. *Global Bioethics*. East Lansing: Michigan State University Press.

Ricoeur, P. 2004. *Parcours de la reconnaissance: trois études*. Paris: Editions Stock.

Schüklenk, U., and B. Bello. 2006. Globalization and health. A developing world perspective on ethical and policy issues. In *Globalization and Health. Challenges for Health, Law and Bioethics*, ed. B. Bennett and G.F. Tomossy, 13–25. Dordrecht: Springer. Second Vatican Council, Pastoral Constitution on the Church in the Modern World *Gaudium et Spes*. Retrieved from: www.vatican.va.

Sinaci, M. (2016. The possibility of global bioethics in a globalized world. In M. Micle and C. Mesaros (Eds.), *Communication Today: An Overview from Online Journalism to Applied Philosophy*, trivent Publishing. Available online at: http://trivent-publishing.eu/.
Ten Have, H., ed. 2015. *Bioethics Education in a Global Perspective*. Dordrecht: Springer.
———. 2016. *Global Bioethics. An Introduction*. London/New York: Routledge.
Ten Have, H., and B. Gordijn, eds. 2014. *Handbook of Global Bioethics*. Dordrecht: Springer.
Turoldo, F. 2007. *La globalizzazione della bioetica*. Padova: Gregoriana Libreria Editrice.
UNESCO. 2005. *Universal declaration on bioethics and human rights*. Retrieved from: http://portal.unesco.org
———. 2008. *Bioethics Core Curriculum*. Retrieved from: http://portal.unesco.org

Chapter 5
Ethics Education Needs More than the Four Principles: Bioethics Discourse in a Community of Inquiry

Leonardo D. de Castro and Isidro Manuel C. Valero

Abstract This essay reexamines the four-principle approach to biomedical ethics in the context of ethics education in general and in relation to possible ethics discourse within a community of inquiry in particular. A community of inquiry is the setting for learning and education in philosophy for children. This community enables children to acquire critical thinking and other skills as part of democratic education. The use (or misuse) of the four principles approach tends to contribute to a practice that limits critical thinking skills because of the constraints on the conceptual tools that tend to be used. It has also had the effect of promoting conceptual ambiguity by encouraging the use of limited conceptual molds, thus giving rise to the possibility of multiple interpretations among diverse users, especially in the field of global bioethics. While recognizing the continuing appeal of the approach as a conceptual tool for ethical decision-making the essay brings out the limitations that need to be overcome in order to promote the clarity that the four principles approach is meant to possess.

5.1 Introduction

Bioethics discourse has been highly reliant on a platform provided by the four-principle approach. Efforts to find solutions to problems and dilemmas confronting biomedical practice and research have led decision-makers to look for conceptual tools that are readily accessible, simply formulated, and widely – if not globally – acceptable. Many academics and biomedical practitioners and researchers are satisfied that they have found these tools in the four principles of autonomy, beneficence, non-maleficence and justice. For the most part, these principles have formed a platform for bioethics discourse that has been well utilized and generally appreciated. Although there have been some detractors, the four-principle approach has had, by

L. D. de Castro (✉) · I. M. C. Valero
University of the Philippines, Diliman, Quezon City, Philippines
e-mail: decastro@kssp.upd.edu.ph

H. ten Have (ed.), *Global Education in Bioethics*, Advancing Global Bioethics 10,
https://doi.org/10.1007/978-3-319-78984-2_5

and large, a huge impact on ethical discourse and decision-making in various areas of bioethics.

The promotion of the four principles started with their publication in the *Principles of Biomedical Ethics* by Beauchamp and Childress (1979), which was preceded by the listing of three of the principles in the Belmont Report submitted to the President of the United States by the National Commission for the Protection of Human Subjects of Biomedical and Behavioral Research in 1978 (National Commission 1978). The Report laid out a framework for protecting human subjects in research that consisted of respect for autonomy, beneficence, and justice. The two publications provided accessible tools for analytical work in biomedicine that appealed to biomedical professionals, researchers and others who felt the need to address emerging ethical issues relating to the rapid growth of biomedical science and technology.

As biomedical science and technology continuously produced new but oftentimes contentious ways of dealing with medical conditions that affected people across the globe, many felt pressured to come up with globally acceptable ways of arriving at decisions regarding ethical issues that arose. The four-principle approach was seen to address the emerging ethical issues globally – i.e., in a way that reflects consensus among people of diverse nationalities or coming from different social, cultural or economic backgrounds. The authors have repeatedly manifested this objective and Beauchamp has said he takes "these principles to be universally valid norms that warrant us in making intercultural and crosscultural judgments about moral depravity, morally misguided beliefs, savage cruelty, and other moral failures (Beauchamp 2003, p. 269). They have also used the term "common morality" to refer to:

> the set of norms shared by all persons committed to morality. The common morality is not merely a morality, in contrast to other moralities. The common morality is applicable to all persons in all places, and we rightly judge all human conduct by its standards. (Beauchamp and Childress 2009, p. 3)

Gillon has firmly supported this view, expressing confidence that none of the four principles can be rejected by anyone and that altogether, these principles constituted the "elements of [a] common moral language and a basic moral commitment for health care ethics that ... can be shared by all regardless of their background" (Gillon 1994, p. 333). He has gone even further to say that:

> the four principles should also be thought of as four moral nucleotides that constitute the moral DNA – capable alone or in combination, of explaining and justifying all the substantive and moral norms of health care ethics and I suspect of ethics in general (Gillon 2003, p. 308).

Indeed, appreciation of the significance of the four principles has remained widespread, with quite a number of professionals highlighting their usefulness, for example, in the area of bioethics mediation. We highlight this area because we want to call attention to a kind of deliberative exercise that involves several parties taking initially disparate positions and is something for which we consider the notion of a community of inquiry – as discussed below – very appropriate. Dubler uses the term "principled resolution" as a fulcrum for resolving or otherwise managing disagreements in the healthcare setting that arise among medical providers, patients, or members of a patient's family (Dubler 2011, p. 177). Clarifying that a principled resolution is a consensus about an action plan that falls within clearly accepted ethical principles, she cites Beauchamp and Childress while observing that the arcane writings of John Stuart Mill and Immanuel Kant have been modernized by the new field of bioethics and, in the process, have given rise to the ethical tenets that include patient autonomy, beneficence, non-malfeasance, and distributive justice (Dubler 2011, p. 178).

Another writer elaborates on the dependence of ethical mediation procedures on the four principles:

> Principles that are assumed as the commonly shared set of norms in bioethics correspond to the four-principle formulation of Beauchamp and Childress … The principles are not just a set of principles: they are *the* principles of medical ethics that apply universally to all morally serious individuals. (Emran 2015, p. 376)

We consider as especially significant the iteration of the view held by Beauchamp and Childress that the four principles are not just any principles of morality but are *the* principles that apply universally. As we argue in this paper, this has a seriously limiting effect on the flourishing of bioethics discourse and the broad objectives of bioethics education. Our argument is not anchored on the criticism that the four-principle approach constitutes an unacceptably inflexible method for arriving at solutions to complex ethical dilemmas. And like Macklin, we take the position that the approach is not rendered entirely useless by its inability to provide clear and unequivocal answers to all moral quandaries that confront us (Macklin 2003, p.275). The four principles constitute a starting point for bioethical deliberation that can facilitate coverage of important considerations and key issues. But we also have to be constantly reminded that we need to go beyond this starting point for reasons that include the inability of the principles to "adequately provide for all of the many roles which health care providers perform and all of the contexts in which they find themselves, as the dilemmas associated with (for example) management and interdisciplinary working might suggest" (Huxtable 2013, p. 42).

Our reasons for taking a critical position with respect to the four-principle approach to bioethics relate to John Harris' view that the four principles are neither the beginning nor end of ethical reflection and that the use of the four principles leads to a sterile bioethics and uniformity of thought among those faced with situations requiring bioethical reflection and decision-making (Harris 2003, p. 303). While acknowledging that many supporters have found the principles to be a sufficient and useful way of approaching ethics, we feel the need to be mindful of the suppression of conceptual fertility and the rejection of the richness and diversity of ethical discourse. We fear that the excessive dependence on the four principles that has been cultivated and promoted tends to create an atmosphere of compliance rather than of understanding and creativity.

As we proceed, we explore the implications of a four-principle approach for bioethics teaching, with specific focus on teaching children. We understand that this is not the usual context in which debate concerning the four principles has taken place but we are confident that this exploration will help us understand the shortcomings and failings of the approach. In order to clarify the context that we have in mind, we examine in detail the origins and objectives of promoting philosophical inquiry with children and exploring this in relation to bioethics teaching.

5.2 The Community of Inquiry and Philosophy for Children

"Philosophy for Children" was created by Matthew Lipman in the latter part of the 1960s in a conscious effort to include philosophy in both the elementary and high school academic curricula. Initially, Lipman envisioned Philosophy for Children or "P4C" as an instrument for improving the reasoning skills of both elementary and high school students by weaving fundamental principles of Formal Logic into philosophical novels. Lipman's own philosophical novels were designed to promote two distinct objectives. First, the main characters of these stories exhibit thinking skills associated with proper reasoning. Initially, the reader is expected to notice that the conversation among the main characters in each novel strictly conforms to the standards of formal logic. Later the reader is expected to assimilate the same skills displayed by the characters in their own conversations. Second, the same characters, through the same conversation, come to a realization that there are limits to the strict application of logical norms in expressing their ideas and proving their conclusions. Consequently, they explore alternative ways of expressing their ideas while manifesting critical, creative and caring thinking skills in order to adequately capture and communicate the multiple layers of meaning that a single idea can have (Lipman 2003, p. 200). The aforementioned objectives are clearly demonstrated in P4C philosophical novels such as *Harry Stottlemeier's Discovery* (Lipman 1974) and *Suki* (Lipman 1978). The principal characters in the plays fully understand the need to conform to prevailing logical norms in both expressing and communicating their ideas clearly and distinctively. However, they also realize that the same norms can severely limit the ideas they want to express and communicate. Consequently, they recognize the importance of sustaining their conversation in order to explore alternative ways of expressing and communicating ideas. We hold that this approach effectively serves as a critical commentary on four principle deliberations on bioethics issues. Taking the four principles as the norm in bioethics discourse can have the effect of severely limiting the ideas that participants can express and communicate. Sustaining bioethics conversation requires the freedom to explore the rich range of additional principles and values without reducing them exclusively to the four principles mould. Enforcing the use of the four principles as the norm for bioethical analysis takes away from those engaged in the process a raw but rich armamentarium of values and other principles that are part of their ethics vocabulary on the pretext that this can already be accounted for by the reduction into one or more of the four principles.

In the Philosophy for Children program, exploring alternative ways of expressing and communicating ideas takes place within the so-called "community of inquiry". A community of inquiry is characterized by dialogue fashioned collaboratively out of the reasoned contribution of all the participants (Sharp 1991, p. 31). The aim is to produce better thinkers and more caring members of society who can tolerate differences at the same time that they submit conflicts to reasonable scrutiny (Turgeon 1998). Children can be educated to identify and agree upon a procedural conception of what it is to reason well. Within an environment characterized by trust and open inquiry they can learn to reason together in trying to arrive at a balanced, humane conception of how to live well. They can also develop a more thoughtful tolerance of the diversity of perspectives that individuals have regarding the topic (Sharp 1987, p. 39).

The teacher who monitors the logical procedures philosophically becomes an equal member of the community, becoming engaged in philosophical deliberations without being perceived to be superior to the students. Following inquiry where it leads, students learn to object to weak reasoning and build on strong reasoning. While accepting the responsibility of making their contributions within the context of the community they accept their dependence on others and also respect others' perspectives. They collaboratively engage in self-correction when necessary and take pride in the accomplishments of the group as well as of oneself. Thus, "in the process, they practice the art of making good judgments within the context of dialogue and communal inquiry" (Sharp 1991, p. 31). As we examine this description of what is supposed to take place in a community of inquiry we are reminded of Harris' observation that bioethics is concerned not so much with identifying the principles or values that ought to apply in a given situation but with "analysing the arguments that are so often already in play and which present themselves as offering solutions in one direction or another" (Harris 2003, p. 303). Taking the cue from the community of inquiry, we can say that bioethics is concerned not so much with identifying which of the four principles ought to apply but with ensuring that the participants perform their roles with the ability to make better distinctions, to recognize underlying assumptions, to distinguish better from worse reasons, to think consistently and comprehensively, to criticize one's own goals and others', and to criticize one's own thinking as well as the thinking of others (Sharp 1987, p. 39). Bearing in mind the important critical process that is thus characterized, we find it difficult to see how the "reasoned contribution of all the participants" can be encouraged to flourish if the justifications that can be invoked were exclusively focused on the four principles.

5.3 Conceptual Apparatus of the Community of Inquiry: Five Aspects

In order to further develop the idea, Sharp identifies five (5) aspects that constitute the conceptual apparatus of the community of inquiry, namely (1) cognitive, (2) social, (3) psychological, (4) moral and (5) political. The conceptual apparatus helps clarify the objectives and characteristics of the community of inquiry and specifies the skill sets students are expected to acquire in order to satisfy the objectives (Sharp 1991, p. 31). Appreciating this conceptual apparatus helps us to understand the program's objectives that we also deem important for bioethics discourse because we see the discourse as taking place within a community of inquiry that is no different from that which provides the setting for P4C. Furthermore, we see the skill sets students are expected to acquire in a community of inquiry to be the same skill sets required for bioethical reasoning. An elaboration of the five aspects enables us to see how discourse of the kind promoted within a community of inquiry can be expected to contribute to bioethical deliberation.

The cognitive aspects of the community of inquiry involve a commitment to follow prevailing logical standards in expressing and communicating ideas. This commitment is supposed to be indicated by the following skills: giving and asking for good reasons, making distinctions and connections, making valid inferences, recognizing logical fallacies, giving counter examples, discovering assumptions, using and recognizing criteria, inferring consequences, defining concepts, judging well, standardizing using good analogies, and offering alternative points of view (Sharp 1991, p. 31–32).

The social aspects of the community of inquiry have to do with the production of knowledge as a social activity. As students reflect on their own personal experiences they are obliged to contribute to the conversation either by listening to, or questioning the validity of the arguments of the other members of the community of inquiry in order to uncover multiple layers of meaning attached to a common idea. If they deliberately choose not to participate in the conversation the quality of the ideas produced diminishes and the outcome may be harmful to the community of inquiry as a whole. Correspondingly, the pertinent skills include listening to one another, supporting one another by amplifying and corroborating their views, submitting one's views of others to critical inquiry, giving reasons to support another's view even if one does not agree, taking one another's ideas seriously by encouraging one another to speak, and speaking when one feels the responsibility of getting the dialogue back on track (Sharp 1991, p. 32).

The psychological aspects of the community of inquiry complement the social aspects. While the production of ideas is a "social" affair, the social aspects of the community of inquiry do not minimize the important role each student plays in the production of ideas. Each individual student is obliged to actively participate in the communal conversation in order to produce the 'best" idea for all the members of the community. However, this same obligation compels all students to constantly re-examine their points of view and reformulate their arguments in order to filter out bias or prejudice from the production of ideas. Thus, not only does the communal conversation benefit all the members of the community; it is also instrumental in helping individual students broaden their appreciation of the significance other members attach to the same idea. The set of skills needed to promote this objective includes putting one's ego in perspective, disciplining self interest, transforming oneself, and refraining from engaging in monologues that do not call for a response (Sharp 1991, p. 32–33).

The moral aspects of the community of inquiry derive from the assumption that education aims to fashion persons of responsibility and integrity, i.e., persons of moral character who are capable of making wise judgments about what is right and wrong (Sharp 1991, p. 34). The assumption entails that participants have to make a "commitment to stay with the group" and that the "community of inquiry constitutes… a way of acting on the world" (Sharp 1991, p. 35).

The political aspects of the community of inquiry are realized when its members make a commitment to "freedom, open debate, pluralism, self-government and democracy,. .. [which can only be achieved in the context of] shared understandings and experiences, interusbjective daily practices, a sense of affinity and solidarity, together with all the tacit affective ties that bind people together in a community" (Sharp 1991, p. 36).

Applying these aspects of the community of inquiry to bioethics education involves trying to guide the learners through the process of achieving the skills associated with domocratic participation in a well functioning community of inquiry. The same skills are important for bioethics discourse in general because of the latter's preoccupation with issues that have to do with everyone not only as individual human beings but also as partaking of humanity. Human beings have to be involved in these deliberations about the future of humanity as well as their own individual futures. Such ultimate concerns as the boundary between life and death, the differences between human and non-human life, the significance of humanity, and the medical enhancement of life require democratic discourse such as that which can be expected to take place only in a well functioning community of inquiry as elaborated on above.

Against this backdrop of the conceptual apparatus of the community of inquiry, the limitations of the four-principle approach can be better understood. The application of the four principles in the work place by medical practitioners and health care professionals can be severely restricted without the awareness of the various aspects that constitute the conceptual platform of the community of inquiry. The ethical assessment of a patient's health care choices can easily degenerate into a grocery list tick-off that is more mechanical than truly deliberative, especially because of the emphasis on simplicity and ease of use. When quick decisions have to be made, there is a temptation to neglect values and considerations other than those that are at the heart of the four-principle approach. Some of these values and considerations can actually shape the meanings of the four principles. The cultural background, political affiliation and religious beliefs of patients complicate matters considerably and procedures for discussion anchored on the four principles simply cannot cope. It is possible that the student's restricted ability to consider the aforementioned variables in thinking about these principles is due to the absence of adequate technology used by medical and health care schools to improve on the quality of discussion about them by their students. Building communities of inquiry with the emphasis on creating open spaces inside the classroom conducive for philosophical conversation may be able to satisfactorily accomplish this objective.

The limitations of the four-principle approach are thus more pronounced when one focuses on the idea of bioethics education. Education in the context of a community of inquiry can be understood to have the characteristics mentioned in the preceding paragraphs. The objectives can only be fully achieved within the community of inquiry in a way that is far from being served by the four principles approach. Moreover, these same objectives provide bioethics education with a cultural, moral and political context. By being more specific about the context, the community of inquiry makes the study and application of these bioethical principles less arbitrary and artificial.

Building communities of inquiry can help enrich the discussion of Principles of Biomedical Ethics both quantitatively and qualitatively. Quantitative enrichment involves the acceptance of additional principles and values even if this implies the existence of overlaps and redundancies among the lot. Overlaps and redundancies should not necessarily have to be avoided because they reflect the raw discourse of common morality. The vocabulary in which common morality is imbedded is rich and diverse. The richness and diversity of language reflects the social, political, cultural and religious differences and variety found in common moral discourse. Reducing all variety that is found in common morality into the four principles artificially diminishes the conceptual tools that are at our disposal and forces us to think according to the rigid and limited categories that correspond to those principles.

On the other hand, through the open spaces provided in a community of inquiry, students can engage in meaningful conversations about these principles, and sharpen their awareness of the cultural, political and religious environment that affect a patient's health care choices. Moreover, through a combination of critical, creative and caring thinking skills (multidimensional thinking) used in these conversations such as clarifying concepts by sharpening distinctions, inferring valid conclusions, avoiding and detecting fallacies, using analogies to settle facts, and challenging assumptions, students may be able to expose multiple layers of meaning (Lipman 2003, p. 200) that people – especially patients – attach to the concepts of autonomy, beneficence and non-maleficence. Most importantly, through the same philosophical conversations, medical practitioners and health care professionals will more likely cultivate both empathy and respect for their patient's choices.

5.4 Democracy and Bioethics Education

Lipman describes how the medical profession best benefits from multidimensional thinking cultivated within the community of inquiry and how the latter can benefit evaluative discourse within the related fields of law and medicine:

> It should be evident now why the law and medicine were cited earlier as likely places to look for exemplary instances of critical thinking. Medicine and law both involve the flexible application of principles (criteria) to practice (judgement), extreme sensitivity to the uniqueness of particular cases (context sensitivity), refusal to allow either principles or facts to become Procrustean beds to which the other is to be fitted, and a commitment to tentative, hypothetical, self – correcting procedures as befits a species of inquiry (self – correction). Both judges and doctors recognize the importance of being judicious: of making good

judgements in carrying out of their practice. Law and medicine at their best illustrate what critical thinking can be and ought to be (Lipman 2003, p. 226).

Similarly, bioethics education can greatly benefit from the community of inquiry approach. Institutionalizing democratic methods – such as that embedded in the community of inquiry, with its emphasis on multidimensional thinking – in teaching bioethics will not only enrich the medical community's appreciation of these ethical concepts that directly affect the profession but also improve the practice of the profession itself. Making ethical judgements is an activity integral to the medical and healthcare professions. Learning to process relevant information regarding the ethical or moral aspects of their decisions stands to be enriched within a democratic setting provided by the community of inquiry. Within the community of inquiry medical and healthcare practitioners learn through multi-dimensional thinking skills to listen to the concerns and opinions of the other members of the community, respect the cultural, political and social context that shape their opinions and care for the well being of the other members of the community.

We are fully aware that the four principles approach has not been intended to address the requirements for democratic bioethics education as Beauchamp and Childress have only sought to provide "a framework for identifying and reflecting on moral problems" (2001, p. 15). This essay should be construed less as a critique of the four principles and more as a commentary on those who look to the four principles approach as a complete tool that can provide answers to bioethical quandaries. We hold with Campbell that the four-principle approach has a "tendency to claims greater than it can justify" and that "a diversity of approach in these complex human scenarios" is needed (Campbell 2003, p. 296). Our endorsement of the virtues of a community of inquiry in the P4C context recognizes the complexity of human bioethical scenarios and helps to promote a diversity of approach that shows a confidence in the ability of the critically minded – and equipped – person (and children specifically) to cope without depending on the crutches provided by the four principles.

Moreover, this essay has provided a critical commentary on the view that bioethics discourse is all about providing answers to bioethical quandaries in the fast-paced context of a clinic or a research laboratory. We discard the idea of education as the transmission of knowledge that is unambiguous and unequivocal by teachers who are seen as authoritative sources of knowledge and stand on the reflective paradigm of education that regards learning as the outcome of participation in a community of inquiry guided by teachers who stir students to think about the world and revealing knowledge to be ambiguous and equivocal (Lipman 2003, p. 18–19). It is important to see that according to this paradigm of democratic education members of the community of inquiry are expected to be reflective, open to others' reasons and judicious. The outcome of the educational process is not necessarily the acquisition of information but a grasp of relationships among varying perspectives. While there are truly situations that require quick decisions under time constraints, the best way to deal with such exigencies is not to rely on the four principles as emergency crutches but to prepare well in advance by participating in the pertinent community of inquiry in a simultaneously learning and teaching role. It could well be that in the entire process of inquiring, the relationships that are experienced and the new insights that are gained are more important than any direct answers that may be arrived at in response to the original question. For in going through the process, the original questions may have to be recast or replaced by new ones. It may be difficult to cope freely with these developments without being prepared to discard one's allegiance to the four principles.

Democratic education itself provides students with the necessary skills they will need in order to cope with the complexities of democratic life. In a democracy, people think about and express their ideas in a number of ways. The methods people use to think about themselves and the world they live in is largely shaped by cultural, political and religious factors. Often these same factors prevent people from developing their potentials as individuals and hinder the ability of communities to embrace democracy fully. The community of inquiry itself as a form of democratic education emphasizes the search for alternative ways of thinking in order to strip off multiple layers of meaning people attach to their ideas. Consequently, students learn the democratic values of openness to and respect for other people's ideas and how these same values shape the communities they live in.

5.5 Consequences for Global Bioethics Education

Adopting Philosophy for Children strategies can be advantageous for global bioethics education in a number of distinct ways, some of which we highlight here. First, through the practice of multi- dimensional thinking, both students of and practitioners in the health care profession will be able to develop a sharper awareness of the many conflicting nuances among values and principles upheld by those who have a stake in the bioethical aspects of health care. These nuances mirror the cultural, political and religious experiences of the global community that may either facilitate or obstruct the efficient application of these values and principles. Second, disagreements in the application of these four principles in actual cases largely owed to the aforementioned nuances can be settled within the community of inquiry. Within the community of inquiry each participant is expected to converse with the other members of the same community in order to hurdle the obstacles these nuances may create and subsequently produce a solution to a problem at hand. Third, through the community of inquiry, health care professionals and other stakeholders will hopefully develop a respectful attitude for these subtle cultural, political and religious differences that inevitably animate both the appreciation for and application of ethical values and principles. Cultivating respect for other is a key ingredient of cultural sensitivity which is a fundamental concept of Global Bioethics Education.

5.6 Conclusion

It is incumbent on the global bioethics education community to recognize the importance of cultivating cultural sensitivity among health care students and professionals. The successful application of ethical values and principles depends on the ability of heath care practitioners and stakeholders to be sensitive to the social and cultural factors that influence the decisions that people make regarding their health and well – being. The positivist /grocery list tick off approach is not particularly suited to accomplishing this goal. As an alternative, a community of inquiry approach is designed to enable participants to be socially and culturally sensitive as participants acquire critical thinking and other skills as part of democratic education. The use (or misuse) of the four principles approach tends to contribute to a practice that limits critical thinking skills because of the constraints on the conceptual tools that tend to be used. It has also had the effect of promoting conceptual ambiguity by encouraging the use of limited conceptual molds, thus giving rise to the possibility of multiple interpretations among diverse users especially in the broad scope of global bioethics. While recognizing the continuing appeal of the approach as a conceptual tool for ethical decision-making this essay has taken up limitations that can be addressed through a community of inquiry approach. Global bioethics education can certainly benefit from strategies designed by the Philosophy for Children program.

References

Beauchamp, T.L. 2003. Methods and principles in biomedical ethics. *Journal of Medical Ethics* 29: 269–274.

Beauchamp, T.L., and J.F. Childress. 2009. *Principles of Biomedical Ethics* (1st edition, 1979; 5th edition 2001). New York: Oxford University Press.

Campbell, A.V.C. 2003. The virtues (and vices) of the four principles. *Journal of Medical Ethics* 29 (5): 292–296.

Dubler, N.N. 2011. A 'principled resolution': The fulcrum for bioethics mediation. *Law and Contemporary Problems* 74 (3): 177–200.

Emran, Ahmadi Nasab. 2015. The four-principle formulation of common morality is at the core of bioethics mediation method. *Medicine Health Care and Philosophy* 18: 371. https://doi.org/10.1007/s11019-014-9612-7.

Gillon, R. 1994. The four principles revisited: A reappraisal. In *Principles of Health Care Ethics*, ed. R. Gillon, 319–333. New York: Wiley.

———. 2003. Ethics needs principles – four can encompass the rest – and respect for autonomy should be first among equals. *Journal of Medical Ethics* 29 (5): 307–312.

Harris, J. 2003. In praise of unprincipled ethics. *Journal of Medical Ethics* 29: 303–306.

Huxtable, R. 2013. For and against the four principles, of biomedical ethics. *Clinical Ethics* 8 (2/3): 39–43.

Lipman, M. 1974. *Harry Stotlemeier's Discovery*. Upper Montclair: Institute for the Advancement of Philosophy for Children.

———. 1978. *Suki*. Upper Montclair: Institute for the Advancement of Philosophy for Children.
———. 2003. *Thinking in Education*. 2nd ed. Cambridge: Montclair State University/Cambridge University Press.
Macklin, R. 2003. Applying the four principles. *Journal of Medical Ethics* 29 (5): 275–280.
National Commission for the Protection of Human Subjects of Biomedical and Behavioral Research. 1978. *The Belmont Report: Ethical Principles and Guidelines for the Protection of Human Subjects of Research*. Washington, DC: U.S. Government Printing Office.
Sharp, A.M. 1987. What is a community of inquiry? *Journal of Moral Education.* 16 (1): 37–45. https://doi.org/10.1080/0305724870160104.
———. 1991. The community of inquiry: Education for democracy. *Thinking* 9 (2): 31–37.
Turgeon, W.C. 1998. *Metaphysical Horizons of Philosophy for Children*. Twentieth World Congress of Philosophy: Philosophy and Children. Available: http://www.bu.edu/wcp/Papers/Chil/ChilTurg.htm. Accessed 15 June 2017.

Chapter 6
Inter-culturality and Cultural Competence

Ayesha Ahmad

Abstract Due to a much more closely connected world, there is an accelerated interchange of cultures in the clinical setting. Alongside a pluralism of cultural beliefs for health and illness is a greater effort to value and respect freedoms for thoughts and belief pertaining to differing identities from legal, moral, societal and activist initiatives. When treating culture as part of the clinical consultation, there is a need for conceptual frameworks, communication skills, guidelines and policies to be implemented in a context where there are certain dominant narratives such as scientific medicine and biomedical frameworks of the human body and mind that typically bracket out culture. The encounter of the clinic with culture is an ethical endeavour that requires expertise training and education; thus, the role of inter-culturality and cultural competence is significant for setting precedents in the translation from theory to practice. In this chapter, I discuss challenges for global bioethics in responding to culture in the clinic.

In the crux of contemporary humanitarian crises resulting from conflict, the demands on global bioethics as a resource for analysis and critique about cultural challenges is increasingly fraught, especially when situated in pluralistic religious and historical discourses in regions affected by trauma from gender-based violence as a weapon of war, torture, and vulnerabilities of hazardous migrant journeys for refugees.

As an academic providing Expert Witness reports to support immigration lawyers that represent asylum seeker cases, I am privy to the transcripts of interviews conducted with different sectors involved in the humanitarian response to war and persecution including Home Office officials (United Kingdom) as well as humanitarian actors from both grass roots and international organisations. In my reports, it is necessary to spend significant time deconstructing and re-explaining cultural features of an individual's trauma narrative. Currently, in a report for a young woman

A. Ahmad (✉)
Lecturer in Global Health, St George's University of London, London, UK
e-mail: a.ahmad@ucl.ac.uk

H. ten Have (ed.), *Global Education in Bioethics*, Advancing Global Bioethics 10,
https://doi.org/10.1007/978-3-319-78984-2_6

seeking asylum in the U.K from Iraq Kurdistan, who was tortured as well as her husband, a journalist, against a backdrop of censorship and political corruption, I had to lay out the cultural background of her current vulnerabilities due to a context of an arranged marriage at aged 14, orphaned due to a previous conflict, and illiteracy. The implications for any slippage in understanding the cultural gaps that are responsible for certain characteristics of her narrative considered to be anomalies or nonsensical are in fact a matter of life and death.

Culture, then, is a pressing ethical endeavour and the roots of the efficacy in humanitarian policy and practice are solidly in global bioethics curriculum, which crosses over multiple disciplines, the approaches to how it is taught, and the ways that the content is prioritised. This chapter tries to discuss these aspects in an applied way to highlight the shortcomings of inter-culturality and cultural competence in both global bioethics education and, ultimately, in the development of core and specialist training of health workers.

6.1 Introduction

Following a lecture that I recently gave at a top British institution on gender based violence in conflict as part of a module on 'Conflict, Health, and Humanitarianism' for an MSc in Global Health, I received an anonymous complaint to my supervisor. The feedback about my lecture from a student who identified herself as a woman of colour stated that she was angry that a woman of colour, namely myself, of Muslim identity, portrayed a case example of Yazidi women and the atrocities they have experienced from the movement of Daesh across their region in the Middle East. My example constituted moments that were described as cringe-worthy by the student to hear her lecturer talk of ill-behaviour of Muslim men to a mostly non-Muslim audience and accompanied this reflection with her anxiety that such revealing of violence would contribute to a growing sense of Islamophobia on campus.

As an academic, I reflected on this incident greatly. There did seem to be some form of cultural encounter of sorts between myself and the student albeit unwittingly. Given that I research and lecture on subjects that are highly sensitive, which is in part my reasoning for bringing the deconstruction of their discourse and subsequently their stigmas and silencing into the educational arena, I am acutely aware of potential responses. However, this specific student feedback resonated long after the lecture and beyond other criticisms or reflections that I may have received. In some ways, I felt a pressure to self-censor my work due to the dynamics between my research areas and the identity that is often constructed and imposed around my ethnicity and religious orientation. Considering this experience, I have questioned the role of cultural identity in delivering global bioethics education as a researcher and activist in the field that I lecture and supervise theses on. Could it be the case that the goals that I want to achieve from my work in terms of improving knowledge and awareness and systematising marginalised topics such as inter-culturality and cultural competence into bioethics education and training for global health and humanitarian researchers and practitioners is compromised by my own identity?

Similarly, I have been accused of racism in delivering material that has critiqued frameworks used in Western psychiatry due to using a reference such as Western as a description in this context. The basis of this criticism was that had it been the case that Islam was used to describe a specific body of knowledge there would be a breaking of political and cultural sensitivities, which again was an example of an undisclosed aspect of my identity being assumed and referred to. Although these experiences have been challenging, they have also been the source of insight into the complexities of cultures within an educational institution.

Of course, as many of the students I lecture to and develop curriculum for are going to be future practitioners in humanitarian contexts, the reactions and challenges of integrating inter-culturality and cultural competence in global bioethics education also provides an indication for how such education would be received and reacted to in the form of frameworks and guidelines during training in the field. Thus, I decided to write this chapter on such a topic to try to embed the discussion of culture in global bioethics education into a wider context.

6.2 Cultural Frameworks in Global Bioethics Education and Humanitarianism

Whilst cultural models for health and illness are being called into greater attention by bioethicists to be included in education and research development, there is a need to question and reflect on the ways that cultural variations in clinical decision-making are framed. Terminologies such as global, for example, imply that health can be divided into the local and the global. Adding in the reference of 'global' when discussing bioethics almost implicitly indicates that culture has now entered the framework. Yet, in our globalised world of accelerated communication, migratory journeys especially healthcare worker transitions, and the significant proportion of the world's population being refugees, the 'local' is often the 'global'. Furthermore, bioethicists need to recognise that "culture should not be simply equated with ethnicity" and account for "processes of socialization into professional cultures, organizational cultures, national civic culture, and transnational culture" (Turner 2005, p. 305).

Considering what framework would be sufficient for global bioethics that accounted for culture, Marshall and Koenig (2004, p. 252) asked there be a "single, "global" bioethics" and continuing on, "how might a global bioethics account for profound—and constantly transforming—sources of cultural difference? Can a uniform, global bioethics be relevant cross-culturally"? These are difficult and challenging questions especially when geared towards developing a standardised and universally applicable curriculum not least because they involve core areas relating to humanity and the human condition, but also because the dominant discourse of bioethics is shaped by positivism and premised on the body of knowledge that is known in the 'Western' world as scientific medicine, and subject to accusations of imperialism and colonialism when travelling across borders, or cultural boundaries.

In virtue of its nature, humanitarianism creates a cultural encounter and this is evident on several levels. The dominant scientific medical framework that is exported across boundaries and implemented in contexts that are very different from that where it originated and formed a structure. Furthermore, the humanitarian physician is bracketed from local cultural understandings of the surrounding world. Developing a cultural lens that critiques these nuances is essential for practicing medicine in a global context especially during humanitarian crises. An important area for global bioethics education to develop is an inquiry about what it means to possess a differentiated cultural body in the context of humanitarian action. Understanding the importance of culture in terms of culture representing a phenomenological relationship with the self and surrounding world will also help to identify crucial factors that need to be accounted for whilst establishing standards for cultural competence.

6.3 Medical Education, Bioethics and Culture

Issues of culture relating to health and illness and the doctor-patient relationship are covered in medical education under the umbrella term of bioethics. Culture is thus fundamentally derived as an ethical issue by medical educators for medical practitioners. The reasons why this is the case is because anything other than a medical motive for interfering with the body or the mental health of another individual are viewed as actions motivated by a cultural explanation. Actions, as with individuals, are at times considered as the 'Other'. Translating culture into curriculum is often polarised between right and wrong, or empirically analysed in terms of being good or harmful to health.

Gregg and Saha (2006) argued that the driver for most cultural competence programs in health professions education is to "alleviate barriers to effective health care for immigrants, refugees, and others on the socio-cultural margin". The presumption that cultural competence is required for only those who are not part of the dominant culture within a local health system and discourse is fallacious in the light of the understandings of culture that have been contributed by medical anthropology. Despite a fuller and more robust discussion of the role of culture in shaping our identities and beliefs about health and illness, the clinical space of biomedical healthcare settings remains reluctant to accepting its mode of practicing medicine as a form of culture. Even on an educational level, culture is an apparatus that is only used when trying to treat and reach conformity and compliance in clinical decision-making with the individual of the 'Other' who performs actions 'other' than those within orthodox scientific medicine.

6.4 Culture and Health: 'All in the Mind?'

Anthropologist E.B. Taylor, coined a seminal phrase on the first page of his two-part book volume 'Primitive Culture' during the early 1920s, referring to culture as that "complex whole which includes knowledge, belief, art, morals, law, custom and any other capabilities and habits acquired by man as a member of society". Culture, then, is an encompassing term referring to all aspects of our human condition and lived experiences.

Culture is essential for threading our narratives of our surrounding world, as well as the world beyond our "surrounding worlds". We form perceptions of our discourses and experiences from the vantage point of the bearings of the self in its orientation of time and space. In this sense, culture is essential for providing a coherent world-view and instrumental for our mental health and well-being; for us to feel that we can make sense of the world around us. Health, then, as a fundamental aspect of our human condition, is subject to cultural conceptualizations and interpretations for understanding and producing meaning about illness and other related aspects of our clinical setting such as birth and death.

For the last 10 years, I have been a member of a Clinical Ethics Committee at a major children's hospital in London, United Kingdom. Quite early on, the type of cases that were being received for ethics review were characterised by questions related to culture or religion—sometimes the distinction was made by either the clinicians/parents of the child/patient, yet other times the distinction was blurred by an assortment of factors such as migration, language, or identification of the reasoning for decision-making on the clinical setting as belonging to religious scripture or tradition. Indeed, in this chapter I shall often refer to culture or religion almost interchangeably or appearing as if there is a distinct category for each but also begging the question for how—or most crucially, why—is there often a need to enquire during such ethical analysis whether the body of knowledge being referred to is cultural or religious? The ambivalence towards approaching culture, or rather non-biomedical, understandings related to health and illness in a clinical setting sparked the beginning of my academic career and research focus on culture in clinical contexts.

Given that international humanitarian organizations working in response to conflict and crisis are comprised of healthcare professionals trained in the same medical education as those who practice medicine in non-humanitarian healthcare settings, I began to investigate the translation of cultural concepts and how culture is structured within education and training for humanitarian workers in settings shaped by cross-cultural relationships and dialogues. The results were harsh in their limitations. Cultural competency frameworks captured but did not address the complexity of the need to be culturally sensitive with the contradiction of humanitarian interventions in virtue of conforming to Western-orientated ethics principles directly interfering and reshaping cultural practices in the settings where their short-term projects are situated. Rather, the educational forum addresses cultural practices in terms of those that are identified as being harmful or as constituting a human rights violation. Once situated in a discourse that is structured on universal premises, the cultural elements become neutralised and bracketed from the practice. Female genital mutilation (FGM) for example is primarily critiqued from a health care lens about risks and harms incurred from the practice, and culture, or sometimes religion, is side-lined as the cause for why the practice occurs and has been sustained through generations.

Fundamentally, culture is bracketed from the clinical context. In a recent teaching session to trainee psychiatrists on culture and trauma, there was a reflection on ethnicity and mental health by a psychiatrist of registrar level (in the United Kingdom, registrar level is one below being a consultant, hence relatively senior). The registrar psychiatrist understood that patients from minority groups tend to use non-orthodox explanatory frameworks and patients of English origin use the biomedical model to understand psychiatric illness. He explained how English patients accept the biomedical model of psychiatry because, as he reflected, there is no culture in England. He concluded that English people no longer go to Church and there is no spirituality element to healthcare.

Of course, this is a singular and anecdotal perspective, but what is revealing is that the way that this registrar practices psychiatry stems from this belief, and the implications are vast. When I responded to the view that the patients who bring culture into the clinical setting are from other cultures and suggested that the biomedical framework of understanding is a culture in its entirety, the registrar paused and accepted this point. Yet, despite over 10 years of medical education, an overseas elective, and practicing in a teaching hospital in an extremely diverse city such as London, this was the first time that the body of knowledge on which he rested his diagnoses and treatment on had been deconstructed into a form of culture. This is problematic and raises concerns about the way that other cultural understandings of mental illness would have been approached. For example, if a patient claimed to be possessed by spirits in a way to understand their symptomology, what would be the response? Defining who has knowledge and who has culture is a dangerous enterprise for the patient in the local or the global clinical encounter.

6.5 The Cultural Encounter

Even in the late 1990s, the way that humanitarian organizations conceptualize and treat culture in response to crises was described to be "problematic" (Hyndman 1998). Likely, the problem is traceable to the curriculum level of culture. Culture presents significant anomalies in the ways that it can be formatted and taught compared to other aspects of humanitarian action such as research ethics during disasters or ethical issues in health management during an intervention.

Culture in contemporary times is arguably best described as being everywhere and nowhere. Political, legal, and health discourses on local, national, and global levels are compulsively debating about culture in the guise of integration, assimilation, and a polarizing rhetoric on migration. Situated in the heart of these discussions is a peculiar notion of culture that has been shaped and formed into an object being passed around the hands of those who are considered to carry culture. The peculiarity is in the way that culture is both unifying and dividing; groups create a collective relationship in terms of a cultural identity, and in turn culture is the medium for drawing the line of where another group begins. Culture, then, is a normative judgment of differentiation of another individual, practice, belief or body of knowledge that is used to form ideas and concepts about health and illness.

Culture is also seen as a vessel, transporting objects of different beliefs, ideologies, practices, and customs and when received by a different land, language, religion, or ethnicity, for example, across geographical divides and borders, normative analyses of such differences impact on the identity of an individual who is perceived to be of another culture as the 'Other'. "Travelling cultures" (Ahmad 2014, p. 68) represent a pluralism of values that exist within a multi-cultural society and challenges the treatment of the body as a universal structure. Different ways of conceptualizing the body in the context of health and illness create different narratives and forms of languages to communicate as a patient to a clinician and vice versa. The global culture of health and illness highlights the need for humanitarian health efforts to be prepared for the cultural encounter in the clinical setting.

The need for cultural competence has indeed been recognised as "global interaction and cultural diversity becomes more prominent" (Chang 2007, p. 187). This is interesting; why is it the case that when there is diversity the need for understanding culture increases? Understanding our cultural navigation has a fundamental role in understanding the way that we perceive our identities and reflections of both ourselves and the worlds surrounding us. We inhabit many worlds; our existential reflections, our families and loved ones, our institutions, our nations, and the places and spaces where our modes of communication orientate us to. Within these worlds are cultures of multiple ontologies. Thus, the idea of culture is inherently untraceable yet conceptualized as an object that can be captured, deciphered and translated.

Cecil Helman explains an understanding of culture based on his experience of a medical doctor trained in Western biomedicine and practicing as a General Practitioner in South Africa as well as being a medical anthropologist. Helman's work focused on the "widening gap between medicine and lay perspectives on ill-health. That is, between 'disease'—the biomedical model with emphasis on physiological data—and 'illness', the subjective experiences of the patient and the meanings that are given to such experiences" (Helman 1991, p. 1081). He also describes how "over the past century, most branches of medicine have made impressive advances in health care. However, certain aspects of the modern medical model limit the efficacy with which it can deal with contemporary health issues in different cultures" (Helman 1991, p. 1080). Helman's argument identifies a systematic contrast between health systems and beliefs premised on the biomedical model, and other modes and forms of practicing treatment and healing.

Helman's work was situated against the backdrop of Arthur Kleinman's writings such as *Patients and Healers in the Context of Culture* (1980) and *The Illness Narratives: Suffering, Healing, and the Human Condition* (1988). Kleinman's research as a psychiatrist explored the experience of mental illness through the lens of a medical anthropologist. Kleinman observed a fundamental distinction between disease and illness; he used the word illness to refer to the "innately human experience of symptoms and suffering" (Kleinman 1988, p. 3). In turn, the experience of illness refers to how the sick person and the members of the family or wider social network perceive, live with, and respond to symptoms and disability. Moreover, illness is a lived experience that exists beyond the primary advent of symptoms, distress, or mental and physical suffering. We can contrast the continuum of illness to the emergency and acute intervention of humanitarian action. The doctor-patient relationship as a clinical encounter with a disease will be temporary. As a cultural encounter with illness, though, the doctor-patient relationship requires a different understanding and ethical framework to accommodate the nuances of a lived experience.

6.6 Culture and Trauma: The Ethics of Diagnosing Suffering

Mental health has attracted very little in terms of funding as part of a humanitarian intervention (Silove et al. 2000). This is despite a review of the Global Burden of Diseases, Injuries and Risk Factors Study 2010 that found that mental and substance use disorders accounted for 183.9 million disability-adjusted life years (DALYs) (Whiteford et al. 2013). Hence the call for countries to prioritize and develop prevention of and treatment of mental and substance use disorders to improve population health and create sustainable health care systems. Nevertheless, there are global inequalities regarding mental health. The treatment gap in mental health is significantly vast: it is estimated that 80% of individuals living with a diagnosable mental health disorder live in low income countries that have only 10% or less of global mental health resources (Jacob and Patel 2014).

Prior to or during a humanitarian crisis, affected individuals and populations may have not had access to mental health services. There are other substantial and culturally contextualized barriers to mental health treatment that the global mental health movement is challenged by. For example, Jacob and Patel (2014) point out that the "vast differences in settings, patient populations, and perspectives between psychiatrists and primary health-care professionals demand caution in the translation of specialist concepts and classifications for use in primary care". This concern is significant since 90% of mental health care worldwide is delivered by non-specialist health professionals. Mental health is heavily stigmatized globally and is not necessarily included in medical education. Due to these factors, some countries such as Sierra Leone only have one psychiatrist practicing in the country.

The translation of terminology related to mental health is a cultural variable. Whereas psychiatrists privilege biomedical evidencing and referencing of disorders, mental illness is interpreted and considered by primary care workers in low income countries as "dimensions of distress" and in light of the earlier discussions from Kleinman, "primary care workers seem to be uncomfortable with the use of the notion of mental disorder, with its disease halo, which sidesteps the disease-illness dichotomy while attempting to encompass both disease and distress" (Jacob and Patel 2014, p. 1433).

6.7 The Global Mental Health Debate: Colonialism or Culture?

Global mental health (GMH) is one aspect of global and humanitarian health that is attached to a side-by-side address of the cultural bearings that shape the understanding and experience of mental illness in different cultural contexts, namely where there is a lack of Western psychiatric resources and mental health system infrastructure. Generally, though, much work is needed to bring culture into the forefront of global bioethics and that the need to consider culture on the field during humanitarian crises is more than a matter of sensitivity and appropriateness; it is essential for effective dialogue and response to health needs.

The field of GMH is evolving and thus the involvement for global bioethics education is essential for the progression of the way mental health is conceptualized and implemented across the globe, most notably in low income countries that have under-developed mental health infrastructure and resources. In *Towards a New Architecture for Global Mental Health* Laurence Kirmayer and Duncan Pedersen (2014, p. 759) address a significant issue of types of knowledge that can inform researchers about outcomes of mental health interventions, for example, in terms of symptom reduction and behaviour change (Bemme and D'Souza 2012).

The development of mental health education and resources globally has a compelling argument—the health injustices linked to mental health such as the treatment gap between individuals who are in need to mental health resources and individuals who have access to available mental health resources or evidenced human rights violations such as chaining or prayer camps where forced exorcisms involving torture occur are understood to be remedied by establishing mental health policy and practice. A contemporary view is that the "idea of affirmative mental health rights can fundamentally advance the dignity and welfare of persons with mental disabilities" (Gostin and Gable 2004, p. 9). Yet the focus on human rights during global bioethics discourse has been prioritised over understanding and critiquing the cultural phenomenologies in which the experience of distress is situated. Summerfield points out that:

> It is telling that published studies of non-Western populations often refer to participant's "limited knowledge of mental disorders", their lack of "mental health literacy", or the need to "teach" health workers and the people they serve about mental health. Here Western psychological discourse is setting out to instruct, regulate, and modernise, presenting as definitive the contemporary Western way of being a person. It is unclear why this should be good for mental health in Africa or Asia. This is medical imperialism, similar to the marginalisation of indigenous knowledge systems in the colonial era, and is generally to the advantage of local populations (Summerfield 2008, p. 993).

A form of global bioethics education that does not account for pluralism of discourses other than Western psychological discourse is creating a vulnerability for responses to mental health. Humanitarian crises are increasingly vulnerable due to a lack of existing locally developed mental health resources and the exposure to suffering. Bhugra (2004) has argued that clinicians practicing mental health in humanitarian settings as well as with migrants fleeing from war-torn areas need to be aware of coping strategies as well as resilience among migrants to avoid interpreting their experiences of suffering as mental illness. In other words, alongside distress, there may also be positive ways that an individual/s developed to respond to their life situation.

Humanitarianism is a response to contexts experiencing crises related to geographical, (geo)political, and social factors. Global bioethics education plays a fundamental role in developing frameworks around health care. Multi-sectorial approaches to responding to healthcare needs are advocated by significant educational and advocacy organisations such as the World Health Organization or the

International Committee for the Red Cross. Yet, the fundamental problem remains that culture is objectified and subject to categorisation rather than being understood on the experiential level. Suffering, though, and its accompanying narratives offers a space to translate different meanings and cultural discourses into ethics frameworks and avoid the perils of such cultural reductionism.

6.8 Post-Traumatic Stress Disorder (PTSD): The Normativity of Suffering

Diagnosing Post-Traumatic Stress Disorder (PTSD) is a normative intervention, which is juxtaposed in the global mental health movement and transcultural psychiatry debate between different meanings of suffering. PTSD is a psychiatric condition that has been subjected to and examined in terms of its cultural meaning. Trauma is a form of suffering and in humanitarian crises, diagnosing PTSD is one of the main focuses for mental health interventions given the prevalence of bearing witness to or experiencing traumatic events. Derek Summerfield speaks of PTSD in terms of personhood arguing that "underpinning the constructs of, say, "mental health" or "trauma" is the concept of a person" (Summerfield 2004, p. 234). Understanding cultural systems and the varying phenomenology of suffering is vital for understanding trauma, yet is bracketed from biomedical frameworks of mental health. Trauma and culture have an important relationship since "traumatic experiences are part of the life cycle, universal in manifestation and occurrence, and typically demand a response from culture in terms of healing, treatment, interventions, counselling, and medical care" (Wilson 2007, p. 3).

The example of trauma, then, provides a significant challenge for biomedical responses to trauma recovery. Without an understanding of the meaning of culture in formulating the ways individuals internalise their surrounding world, policies and frameworks addressing traumatization will be at risk of failing. Culture is both a shared and internal mode of navigating and making sense of lived experiences, yet trauma is the disruption and disintegration of such existing world views. Judith Herman has described traumatic events to be "extraordinary, not because they occur rarely, but rather because they overwhelm the ordinary adaptations to life" (Herman 1992 p. 377).

The loss of familiarity and the non-alliance of cultural symbolisms with what has happened to an individual during a trauma involves an experience that is "beyond the psychological dimension of suffering it involves" and "suggests a certain paradox: that the most direct seeing of a violent event may occur as an absolute inability to know it; that immediacy, paradoxically, may take the form of belatedness" (Caruth 2016 p. 94). The violence referred to by Caruth pertains to the rupturing of an individual's narrative including a collective shared narrative in certain traumas such as a disaster. Trauma, then performs through rupture, paradox, semantic reduction, deconstruction of identity and conflicts with the ability to narrative. Trauma therapeutic interventions such as Narrative Exposure Therapy are dependent on the recounting of a trauma narrative and mark the recovery from trauma by assessing the extent of the reconstruction a pre-ruptured narrative. However, there are significant cross-cultural conceptions of narrative especially in contexts of oral story-telling traditions.

6.9 Understanding Culture: Barriers of Cultural Conceptions

Due to a polarised view of culture versus non-culture the trajectory that unfolds leads to an essentialist view of being cultural. There are two aspects here to consider; being cultural in terms of who is the cultural victim—the individual who is perceived as being from a culture, and being cultural in the sense of conforming to dominant terms such as appropriateness or sensitivity or competency. In responding to catastrophic events, health care providers are confronted with frameworks of ethics of care that juggle different categories of culture. An ethics of care that recognises the need to manage complex cultural considerations by displaying cultural sensitivity overrides any adequate analysis of the phenomena that constitutes culture.

A fundamental component of global bioethics education for culture needs to be an in-depth exploration of values held by the student and future health care practitioner. In humanitarian contexts, the practitioner's spiritual distress is brought to awareness in organisational polices and managed in terms of its psychological vulnerabilities; however, culture remains as an indestructible object. Being culturally sensitive or culturally aware is disproportionally referring to cultures that are the 'Other', rather than one's own, and the sensitivity lacks description about what this entails.

6.10 Culture Clash: When Cultures Are Harmful

Grounding an analysis of culture in the educational sphere of global health and humanitarian practice requires an exposition and dissemination of research into the dialogue and critique of healthcare. Health, in the humanitarian context, is situated in a multiplicity of frameworks due to the need to address the social determinants of the crisis, which in turn requires a multi-sectorial approach. Global bioethics education will need to address and learn the languages of other fields of inquiry. One area that needs significant overlap and conversation with global bioethics is disaster research. Developing disaster risk reduction, management and health responses has undergone a shift in its own cultural formation of a humanitarian crises. As disasters have gradually become to be understood as being the acting out of pre-existing social phenomenon, there has been a move from a previously reductionist view of a disaster as a specific event affecting a specific group of people for a specific period. Theorists working on gender have subsequently gained a presence in disaster research and gender lens and discussions about vulnerable groups are now embedded in diagnosing and responding to disasters. Maureen Fordham's (1998, 2003) work has argued for recognition of different forms of a disaster other than the disastrous event providing an evidence basis and gender framework of understanding a disaster. In this sense, Fordham has identified that for women and girls the disaster occurs at a later phase after the initial effects of the disaster have been responded to. Of course, this has serious implications for considering how to shape humanitarian interventions and respond to ethical issues.

Following a recent trip to Afghanistan to analyse interviews conducted in Safe Houses for women who have experienced gender-based violence, I encountered a complex culture within the vast presence of humanitarian aid workers representing international non-governmental organizations within the country. During an interview with an Afghan doctor, he explained the frustration of working on a U.S. funded program to prevent gender-based violence. Since the basis for gender-based violence is predominantly understood as a lack of gender equality, such prevention of gender-based violence programs in Afghanistan are marketed in exactly this way, namely, as ways to develop gender equality. However, in a context where the society is highly gendered as the result of an extremely complex range of factors including nearly 40 years of conflict, women are psychologically and socially confined to their homes, and education is not viewed as necessary for young girls. Gender equality programs have been interpreted by communities as attempts by foreigners to divide society and separate families by encouraging women to gain a greater presence in the workplace and for girls to go to school. This tension, the Afghan doctor suggested, could be avoided if the programs were referred to using more culturally acceptable concepts and terminologies rather than frameworks that originate from and are orientated to the 'West'. On the global bioethics education spectrum, it is the violence towards women and girls that is viewed as the ethical issue, and not the conceptualization of the principles and practice of programmes designed to prevent violence against women and girls.

Another way that culture clashed in a harmful way was the way that cultural customs such as 'baad' or 'baadal', loosely translating to the exchange of a daughter from a family, usually a child, for marriage to settle a feud, have been equated as the cause of violence towards women and girls. Attributing culture to the existence and perpetuation of the mythological rather than to man-made customs creates an elusiveness—what is this thing called culture we are referring to here and where is it located? On a pragmatic level, the transcendence of an action that results in the violation and violence of Afghan women into the cultural sphere removes any accountability and avenues for exposure of the structural and systematic use of violence against women and girls, for example, addressing the presence of war lords in a government who hold strict and extreme ideologies against women that support and fuel the continuation of such customs.

The interviews revealed perpetrators of violence that hold authority positions such as police, doctors, religious leaders, and governmental officials. Clearly, viewing the gender-based violence suffered by the women who gave these interviews despite the risks involved in disclosing their story, as 'culture' will forfeit any attempts to address violence against women and girls from humanitarian health action. Global bioethics education is thus an activist and advocacy tool; a sufficient development and integration of global bioethics into humanitarianism should highlight the live nature of research and humanitarian interventions in shaping cultural narratives that have become infused with and underpinned by violence from other sources such as geopolitical or symptomatic of cultures within humanitarian organisations. Ethics is "something far more profound than a subject of academic study and scholarly discourse" and "bioethics can help lead policy makers towards a more holistic approach to clarifying and finding the most equitable and ethical means to achieve their desired goals" (Fox 2001, p. 33). These are observations that lean to the necessity of the instrumentalization of ethics throughout global discourses—global not just in the sense of geographical identification but in the pluralism of forms of knowledge and conceptualizations of culture.

6.11 Conclusion

Inter-culturality and cultural competence continue to gain momentum in global bioethics. Education about culture and its accompanying ethical issues can be sourced from a variety of disciplines and inter-disciplines, which provides a challenge for ensuring that relevant research and advocacy reaches the attention of bioethicists, and that bioethics gains members from the global community. Humanitarian crises, at the same time, are increasingly complex. In part there is a greater understanding of the nature of a disaster as a socio-cultural body rather than a biological construct, as well as a changing nature of contemporary conflicts being protracted and motivated by fundamentalist ideologies. In turn, the suffering of individuals caught up in humanitarian crises is characterised by violence intended based on identity, such as gender or religion or ethnicity. The experience of cultural orientation and identity is crucial for understanding different stories of suffering. Humanitarian interventions have a significant ethical agency and responsibility on the bearing of values such as equality and justice in recovery for both individuals and societies. The frameworks for these values need careful and robust ethical analysis and this includes epistemological inquiry as well as the deconstruction of ethical issues that arise from the differentiation of the cultural body. Global bioethics education has the potential for a systematic intervention as part of humanitarian discourse and response to health needs and suffering. Serious attention to our own understandings of culture and our cultural values, perceptions, and meanings are vital for a balanced curriculum especially when translated globally.

References

Ahmad, A. 2014. Do motives matter in male circumcision? Conscientious objection against the circumcision of a Muslim child with a blood disorder. *Bioethics* 28 (2): 67–75.

Bemme, D., and N. D'souza. 2012. Global mental health and its discontents. *Science, Medicine and Anthropology* 51(6): 1.

Bhugra, D. 2004. Migration and mental health. *Acta Psychiatrica Scandinavica* 109 (4): 243–258.

Caruth, C. 2016. *Unclaimed Experience: Trauma, Narrative, and History*. Baltimore: JHU Press.

Chang, W.W. 2007. Cultural competence of international humanitarian workers. *Adult Education Quarterly* 57 (3): 187–204.

Fordham, M.H. 1998. Making women visible in disasters: Problematising the private domain. *Disasters* 22 (2): 126–143.
Fordham, M. 2003. *Gender, Disaster and Development. Natural Disasters and Development in a Globalizing world*, 57–74. London: Routledge.
Fox, M.W. 2001. *Bringing Life to Ethics: Global Bioethics for a Humane Society*. New York: SUNY Press.
Gostin, L.O., and L. Gable. 2004. The human rights of persons with mental disabilities: A global perspective on the application of human rights principles to mental health. *Maryland Law Review* 63 (20): 8–31.
Gregg, J., and S. Saha. 2006. Losing culture on the way to competence: The use and misuse of culture in medical education. *Academic Medicine.* 81 (6): 542–547.
Helman, C.G. 1991. Limits of biomedical explanation. *The Lancet* 337 (8749): 1080–1083.
Herman, J.L. 1992. Complex PTSD: A syndrome in survivors of prolonged and repeated trauma. *Journal of Traumatic Stress* 5 (3): 377–391.
Hyndman, J. 1998. Managing difference: Gender and culture in humanitarian emergencies. *Gender, Place and Culture: A Journal of Feminist Geography* 5 (3): 241–260.
Jacob, K.S., and V. Patel. 2014. Classification of mental disorders: A global mental health perspective. *The Lancet* 383 (9926): 1433–1435.
Kirmayer, L.J., and D. Pedersen. 2014. Towards a new architecture for global mental health. *Transcultural Psychiatry* 51 (6): 759–776.
Kleinman, A. 1980. *Patients and Healers in the Context of Culture: An Exploration of the Borderland Between Anthropology, Medicine, and Psychiatry*. Vol. 3. Berkeley: University of California Press.
———. 1988. *The Illness Narratives: Suffering, Healing, and the Human Condition*. New York: Basic Books.
Marshall, P., and B. Koenig. 2004. Accounting for culture in a globalized bioethics. *The Journal of Law, Medicine & Ethics* 32 (2): 252–266.
Silove, D., S. Ekblad, and R. Mollica. 2000. The rights of the severely mentally ill in post-conflict societies. *The Lancet* 355 (9214): 1548.
Summerfield, D. 2004. 12 cross-cultural perspectives on the medicalization of human suffering. In *Posttraumatic Stress Disorder: Issues and Controversies*, ed. Gerald M. Rosen, 233. Seattle: Wiley.
———. 2008. How scientifically valid is the knowledge base of global mental health? *BMJ: British Medical Journal* 336 (7651): 992–994.
Turner, L. 2005. From the local to the global: Bioethics and the concept of culture. *Journal of Medicine and Philosophy* 30 (3): 305–320.
Tylor, E.B. 1871. *Primitive Culture*. Vol. 2. Michigan: University of Michigan Press.
Whiteford, H.A., L. Degenhardt, J. Rehm, A.J. Baxter, A.J. Ferrari, H.E. Erskine, F.J. Charlson, R.E. Norman, A.D. Flaxman, N. Johns, and R. Burstein. 2013. Global burden of disease attributable to mental and substance use disorders: Findings from the Global Burden of Disease Study 2010. *The Lancet* 382 (9904): 1575–1586.
Wilson, J.P. 2007. The lens of culture: Theoretical and conceptual perspectives in the assessment of psychological trauma and PTSD. In *Cross-Cultural Assessment of Psychological Trauma and PTSD*, ed. John P. Wilson and Catherine C. So-Kum Tang, 3–30. New York: Springer.

Chapter 7
Resources for Global Ethics Education

Aimee Zellers

Abstract As we continue to grow as global community we need resources that focus specifically on global ethics issues, and available—or lack of available—resources makes it clear that the time has come for ethics to embrace a global perspective. This chapter provides a survey of resources currently available to facilitate educational experiences in global ethics. Practitioners as well as educators will find resources including specific print and digital media as well as databases, repositories, film collections, centers and teaching tools. The chapter concludes with current evidence-based pedagogical practices in ethics education that can be adapted to global ethics education.

7.1 Introduction

Given the extent to which our lives are affected by what goes on around the world, it is time to rethink ethics education. As seen with SARS and Ebola, infectious diseases can be carried from one continent to the next. And as demonstrated by Chernobyl and Fukushima, environmental disasters in one area can have long lasting impact on other peoples and places. In many respects global *is* local.

Furthermore, we have a great deal to gain by expanding our horizons. We see this in Bioethics, for example. Ilhan Ilkilic's work on the impact of culture and traditions on truth-telling calls into question our assumptions around honesty in patient doctor-patient relationships (Ilkilic 2014). Cecilia Wee shows in her work on Confucian Bioethics, that bringing role morality into the discussion of acts of omission vs. commission could reshape the euthanasia debate (Wee 2014). And Peter Omonzejele's analysis of vulnerable populations in Nigeria reveals some fundamental problems with informed consent in human subject research (Omonzejele 2014). Their contributions offer valuable perspectives on important bioethical

A. Zellers (✉)
Carlow University, Pittsburgh, PA, USA
e-mail: adzellers@carlow.edu

H. ten Have (ed.), *Global Education in Bioethics*, Advancing Global Bioethics 10,
https://doi.org/10.1007/978-3-319-78984-2_7

issues. Assumptions shaped by the Western worldview have taken us a long way but there's a lot to be gained by taking a wider—more global—view.

While there are many resources for ethics education from a western perspective there are no comprehensive courses that address global ethics as such. There is some excellent work being done in specific pockets of applied ethics such as global business ethics (Weaver 2001; Rossouw 2011; Choi et al. 2010) and global bioethics (ten Have 2015). Yet, there is a general lack of quality resources that focus primarily on global ethical issues, and available resources makes it clear that the time has come for ethics to embrace a global perspective. It is up to the educator to make connections between the western perspective and plurality of other worldviews. This chapter aims at helping educators make some of those connections by demonstrating where to look for assistance. To accomplish this, this chapter will survey resources available in a global ethics education. Practitioners as well as educators will find resources including specific written (books, Ebooks, articles, etc.) and digital media (films, documentaries, podcasts, etc.) as well as databases and repositories. The chapter will conclude with evidence-based pedagogical practices in ethics education that can be adapted to global ethics education.

7.2 Written Resources

As technological innovations and capacity grow, even our traditional "print resources" are accessed and organized digitally in many cases. This section will address those written sources that can be most easily located online. Section one will identify a number of recent publications and helpful journals. As an exhaustive list of recent publications cannot be provided here, section two will provide excellent options for locating archived and recent publications pertaining to global ethics education.

7.2.1 Journals and Books

In our digital age, new publications seem to arrive on a minute-by-minute basis. The following list of books and reference journals is very narrow in focus, as it is impossible to capture even a brief snapshot of all relevant publications. The books below include only those publications relating to global ethics education that have been published since January 2015. The journals listed below capture only those that exclusively address global ethics education. There are many excellent ethics and applied ethics journals, but they fall outside the scope of this chapter. From the brevity of the lists below, it is clear that there are numerous titles addressing highly relevant issues, but not the issues of global ethics education itself.

Recent Book and Ebook Publications

- *Teaching College Students How to Solve Real-Life Moral Dilemmas: An Ethical Compass for Quarterlifers* by Robert Nash and Jennifer Land (2016)

- *Globalization of Bioethics Education* by Henk ten Have (2015)
- *Ethics, Aesthetics, and Education* by Donald Blumenfled-Jones (2016)
- *Ethics and Diversity in Business Management Education* by Mary Godwyn (2015)
- *The Ethics of Cultural Competence in Higher Education* by Heidi Schnackenberg and Beverly Burnell (2015)

Journal Titles

- *International Journal of Ethics Education* (ISSN Online: 2364-0006) (since 2016)
- *Science and Engineering Ethics* (ISSN Online: 1471-5546) (since 1995)
- *Journal of Academic Ethics* (ISSN Online: 1572-8544) (since 2003)
- *Ethics and Global Politics* (ISSN Online: 1654-6369) (since 2008)
- *Global Virtue Ethics Review* (ISSN Online: 1523-9756) (since 2000)
- *Global Bioethics* (ISSN: 1128-7462 Online ISSN: 1591-7398) (since 1988)

7.2.2 Databases

Databases provide a very important function in that they allow to search of millions of sources simultaneously. This dramatically increases research effectiveness, if the databases are used correctly. Before using any of the databases listed below, users should acquaint themselves with any unique functions or parameters of the database and be able to conduct a basic Boolean search. The databases listed below are excellent resources for global ethics sources. They are organized in descending order from very specific to very broad with regard to content collection and quality. The list includes: The Global Digital Library on Ethics, The Global Ethics Observatory, BELIT, EthxWeb, EthicShare, the Philosopher's Index, PhilPapers, Project MUSE, PubMed, ProQuest, and Google Scholar.

The *Global Digital Library on Ethics* (GlobeEthicsLib) hosted by Globethics.net is a worldwide ethics library and network based in Geneva. In addition to offering access to more than four million full-text sources, the site also provides ways for researchers to connect and collaborate. Currently there are more than 170,000 registered users (Globethics 2016).

The *Global Ethics Observatory* (GEObs) is a system of linked databases focused exclusively on applied ethics and bioethics. It is a United Nations Educational, Scientific and Cultural Organization (UNESCO) network and as such is worldwide in scope and available in six languages. The databases are divided into six categories. The two most relevant for global ethics education include Database 3: Ethics Teaching Programmes and Database 6: Resources in Ethics (UNESCO 2016a).

BELIT, better known as the *Bioethics Literature Database*, which is facilitated by ethicsweb.eu is an initiative of European Ethics Documentation and Information Centers. It provides access to an international network of resources from the

United Kingdom, Germany, American and French databases (ETHICSWEB 2016).

EthxWeb is a database indexed by Georgetown University that consolidates interdisciplinary bioethics material from 1974 to 2009 (Georgetown University 2016b). It is a component of the DigitalGeorgetown interface. The added benefit of this database is that is open source and includes a "Bibliographies of Bioethics Topics" page. This page contains bibliographies for over 40 topic-based searches and they are available in downloadable PDF formats (Georgetown University 2016a).

EthicShare is a website that supports a virtual community that shares research and collaborates on ethics-related issues and topics. It was original designed by the University of Minnesota (UMN) Libraries in conjunction with the UMN Center of Bioethics and UMN Computer Science Department. This project received national funding. The community allows searching the shared research collections, creating group discussions, news reports, and an events calendar (UMN Center for Bioethics and UMN Computer Science Department 2016).

The *Philosopher's Index* is a unique tool for locating philosophical ethics texts and commentaries. It contains more than 600,000 records, originating from 139 countries and it is best accessed through an existing library subscription. In addition the full text database can be search through EBSCOhost (Index 2016).

PhilPapers is an index and bibliography of more than two million sources of the research literature in philosophy. Locating or developing a bibliography for philosophical ethics is made much more accessible with this index (PhilPaper 2016).

Project MUSE is an excellent database for researchers as it makes the resources of 120 publishing companies in the digital humanities and social science content available. It includes only the top university presses including Oxford, Cambridge, MIT, Harvard and Columbia as well as world-renewed academic centers and collections including the Hastings Center, Middle East Institute, and African Books Collective (Project MUSE 2016).

PubMed is a popular database that contains over 26 million citations for biomedical literature from MEDLINE, life science journals, and online books (U.S. National Library of Medicine 2016).

ProQuest Research Library is an online multidisciplinary database featuring a highly diverse set of sources including thesis/dissertations, Ebooks, journals, magazines, etc. This resource is available for purchases a subscription or if an individual has access through an existing library or institution (ProQuest 2016).

When ethics-centered databases fail to yield desired results, there are broader search engines that can be utilized, such as Yahoo.com and Google Scholar. There are important pros and cons, to consider. For example, a disadvantage is that searches often yield a lot of junk websites, blogs, and less-than-credible sources. On the other hand, both Google Scholar and Yahoo have a broad reach and allow users to set up alerts so they can be kept up-to-date on topics of interests. This is an extremely useful feature for ethics education as there are daily news updates on a range of local and international issues.

7.3 Digital Resources

Digital resources refer to a variety of available resources including video, audio, and any combination thereof. This section will review available resources in video format, including documentaries, films, and video repositories, as well as podcasts, online courses, and education materials that can be accessed online.

7.3.1 Documentaries, Feature Films, and Other Video Resources

There are an incredible number of documentaries and feature films that can be adapted to teaching global ethics, even if they do not focus on ethics or global ethics education exclusively. In fact, there are several centers and academic programs, including the University of Miami Ethics Program (University of Miami Ethics Program 2016) and the Kenan Institute for ethics at Duke University (The Kenan Institute for Ethics 2016) that have started utilizing ethics film series as a teaching tool. This section on video resources includes two categories. The first includes documentaries and feature films that address human rights issues. The second category includes organizations that keep updated lists or archives of documentaries and feature films that address ethics and global ethics issues.

The following are international films addressing global human rights along with the corresponding country of origin or focus. Documentaries and docudramas like these are valuable in teach global perspectives and relating them to concrete ethics issues. Ten excellent documentaries include: *What Tomorrow Brings* (Afghanistan); *Burden of Peace* (Guatemala); *The Wanted 18* (Palestine); *No Land's Song* (Iran); *Cartel Land* (Mexico and the United States); *The Trials of Spring* (Tunisia, Egypt, Libya, Yemen, Bahrain and Syria); *The Look of Silence* (Indonesia); *Life is Sacred* (Colombia); *Beats of the Antonov* (Sudan); and, *The Black Panthers: Vanguard of the Revolution* (United States).

There are several of organizations that categorize and archive media so it can be easily searched. There three organizations addressed below contain lists of existing films specific to ethics and global issues: the Carnegie Council for Ethics in International Affairs, the Library of Congress, and Ethics for Everyone. In addition to sites that specialize in global ethics, there are also sites that specialize in making quality documentaries available to the public. There are few excellent sites that facilitate this, and all of the following are mostly free of charge. These sites organize documentaries in general and are not exclusively geared to the topic of global ethics education, or ethics at all. The following sites will be addressed below: Global Oneness Project, TED Talks, What's the Big Idea?, Documentary Heaven, and YouTube.

Carnegie Council for Ethics in International Affairs provides an extensive list of feature films and documentaries pertaining to ethical issues. The list is organized into topics such as oppression and conflict, social and environmental issues, war and peace, genocide, etc. In addition, there is also a "full review" section for a number of films listed. A full review includes an overview of the film, details on the ethical issues addressed, discussion questions, and related sources (Carnegie Council for Ethics in International Affairs 2015).

Library of Congress is one of the most useful sources for conducing broad searches on movies and films, since it is the largest library in the world. For example, a simple search in the Film and Video portion of the online library using the search term "global ethics" yielded 52 results (including films, collections, and series), 48 of which are available online. Titles such as *Translating Africa in Global Contexts*, and *Humanity in Crisis: Ethical Responsibility to People Displaced by War* are easily located. If the search term is broadened to "ethics" more than 290 results appear, 123 of which are available online. This search tool allows users to focus their results by country/location, topic, and digital format (Library of Congress 2016).

Ethics for Everyone is a site that seeks to organize popular culture films around traditional ethics themes such as abortion, euthanasia, identity, racism, and others (Ethics for Everyone 2016). While the list is not nearly as extensive as those of the Carnegie Council or Library of Congress, the organization by traditional themes may be a useful place for educators to start.

The Global Oneness Project offers free multicultural stories and new content is added monthly. Their films have been featured on *National Geographic*, PBS, *The Atlantic*, *The New York Times*, *The New Yorker*, and at the Smithsonian. In addition to providing high quality productions in film, photo essays, and articles, the Project also provides multimedia lesson plans that are free and available online. The lesson plans are easily adaptable to both high school and college classrooms and learning contexts (Global Oneness Project 2016).

TED Talks have gained massive popularity in higher education over the past decade. TED is a nonprofit organization aimed at spreading powerful and important ideas through short talks by leading and emerging experts (typically 18 minutes or less). With over 2300 talks archived online and two subgroups dedicated to global issues and morality this proves to be a powerful educational tool. The search function allows users to search by topic, duration, and language (TED 2016).

There are also resources available for young learners. What's the Big Idea? is a site aimed toward introducing middle school students (roughly 10–14 years of age) to philosophical concepts through film. The site provides high quality, age appropriate content along with instructional tools and videos for educators. This initiative is funded by the Massachusetts Foundation for the Humanities, the Northamption Education Foundation, and the American Philosophical Association (Wartenberg and Akeret 2016).

Documentary Heaven is a collection of documentaries that focus on the environment, health, society, and politics. While the list is not exclusively ethics focused, all the topics addressed have inherent ethical dimensions. The documentaries are organized into 39 different themes including art, health, human rights, gangs, and more. The site also provides a comments feature for each of the documentaries in which users can provide reviews or add relevant sources (Documentary Heaven

2016). One negative aspect of this resource is that it allows users to submit content. While that content is screened for appropriateness it may not always be screened for accuracy. As always, educators should screen and thoroughly review all materials before making them available to learners.

YouTube.com, the ubiquitous site that contains a lot of rubbish, can be a fantastic tool if used appropriately. YouTube is user-friendly and is accessible in most countries.[1] However, there are serious concerns about accuracy, credibility of sources and information, search capabilities, and so on. Despite existing limitations, YouTube does provide one element that assists in mitigating these limitations, and that is the YouTube channel feature. A YouTube channel is home page for an individual or organization's account. This means that the content on the channel page is controlled and managed by a specific individual or group, and the public can view any videos or user information that has been uploaded. One need only subscribe to a YouTube channel to view the content. It should be noted that, in some cases, fees do apply, and the individual or organization managing the channel sets these fees. Two examples of a high quality YouTube channels are the "Global Ethics Corner" which is sponsored by the Carnegie Council for Ethics in International Affairs (Carnegie Council for Ethics in International Affairs 2016a), and the "Center for Global Ethics and Politics" sponsored by The Graduate Center at City University of New York (Center for Global Ethics and Politics 2016).

7.3.2 *Podcasts*

Podcasts are digital audio files available for download on a computer or mobile device. There are a number of advantages in using podcasts for educational purposes including enhanced reading events and enhanced listening comprehension which improves literacy (Godsey 2016). Podcasts are a convenient on the go solution for busy working professionals since one can take the lecture, talk, interview or whatever the podcast content might be with them. There are a host of mobile apps available to enhance the listening experience. Currently, some of the best podcast apps available include Overcast (iOS only), Pocket Cast (iOS and Android), Castro (iOS only), Spotify (iOS and Android), Deezer (iOS and Android), and, of course, Podcasts (iOS only). All of these apps can be found in the Apple (iOS) or Android app search function. Each of these apps has a unique interface that allows one to search, download, and organize podcasts. There are also a number of sites that will host an organization or individual's podcast, including iTunes, SoundCloud, Archive.org, Amazon S3, Ourmedia.org, Podomatic, Libsyn, and PodBean. Several of these sites are free of charge so long as uploaded files do not exceed a specified size. Hosting podcasts on a personal or organization website works, however, there are advantages to posting podcasts to a separate podcast host like those mentioned

[1] Over the course of its existence, YouTube has been banned at one time or another by as many as 24 countries.

above. The most notable advantage is that a podcast host will provide listening analytics, as well as help share the podcast with a much wider audience.

iTunes has proven to be an excellent host for organizing and providing access to high quality podcasts. As access to the iPhone is near worldwide, this specific host for podcasts presents a unique global access feature. iTunes is not only available on Apple products but can also be downloaded on to non-Apple products. Universities and academic institutes have and are currently taking advantage of this free software. A simple search of "global ethics" in iTunes yielded podcasts from Oxford University's Uehiro Center for Practical Ethics, College of the Holy Cross's MacFarland Center for Religion, Ethics and Culture at Holy Cross, Wellesely College's Albright Institute for Global Affairs and many more. These downloadable podcasts are great starting points for introducing students to new concepts, case studies, and content from top-tier universities. The iTunes search feature allows users to search by author (institution) as well as theme or topic.

7.3.3 Open Online Courses and Collaborations

In addition to standalone podcasts and themed series, several colleges and universities have made full courses lectures on the topics of philosophical ethics, normative ethics, and applied ethics available in iTunes. These are located in a subgroup of podcasts called iTunesU ("U" stands for university). In iTunesU, the universities with the largest libraries include Liberty University, Harvard University, University of New Orleans, Ohio State University, Open University and Yale University. A simple keyword search in the iTunesU/Podcast page will allow one to search for relevant courses. Several of these courses also come with transcripts of the podcasts to enhance student learning and retention of information; this varies depending up the course and entity offering the course. Not all courses are free.

In addition to courses on iTunesU, there are a number of additional digital tools that have already been developed to enhance global ethics education, one such tool is massive open online courses (MOOCs). MOOCs were at one time thought to be the digital education innovation of the future because it made education essentially free and open to everyone. As with its emergence 8 years ago, this form of online education is still criticized by some academics. Initially, these courses were hosted by universities and most offered for free of charge, now private companies are partnering with universities to organize, archive, update, and market courses (Daniel 2012; Adams et al. 2013). As a result, virtually all now require some type of fee.

Top MOOC providers include Coursera, Udacity, edX, and FutureLearn. All of these function in very similar ways, thus only Coursera will be explored here as an example of how MOOCs can be beneficial to global education. Coursera has had continued success coordinating MOOCs from a variety of universities on a wide range of topics. Anyone can enroll in one of the courses, assuming they pay the course fee (which usually ranges between $29 and $99 USD). All courses are taught by "top instructors from the world's best universities and educational institutions"

(Coursera 2016). At present Coursera boasts more than 1600 available courses, with 130 specializations. They currently partner with more than 145 universities, including Stanford University, University of Pennsylvania, Johns Hopkins University, and Duke University to reach more than 22 million learners. According to the online course catalogue there are approximately 43 courses that address ethics and approximately 185 courses that address global perspectives (Coursera 2016). Coursera and other similar companies provide a useful service for those who are interested in learning from experts. However, standalone courses and specializations do not carry college course equivalent credits. This means that you cannot take a course certificate from Coursera to a college or university and expect them to add it to your academic transcript. One can, however, add it to a curriculum vitae or LinkedIn profile as professional development (Coursera 2016).

7.3.4 Sample Courses and Educational Material

There are few altruistic organizations that make ethics-related lesson plans, sample syllabi, and other educational resources available free of charge. These are typically non-profit organizations such as United Nations Educational, Scientific and Cultural Organization (UNESCO) and well-established colleges and universities. The following sections will address organizations that provide educational resources including Massachusetts Institute of Technology, Ethics CORE, the Ethics Lab, and UNESCO.

In 2000, Massachusetts Institute of Technology (MIT) proposed initiative called the OpenCourseWare project (OCW). The goals was to make all course materials available online. In seven short years, MIT faculty took the OCW idea and produced an initial publication of nearly the entirety of the MIT undergraduate curriculum. That is approximately1,800 courses in 33 disciplines. This means that course material (i.e. syllabi, readings, assignments, and assessments) are available online—for free. Perfectly stated by Dick K.P. Yue, Professor, MIT School of Engineering: "The idea is simple: to publish all of our course material online and make them widely available to everyone" (About OCW, MIT 2016a). As of 2014, 2250 courses had been published and the main page has been viewed more than 170 million times. This is free resource that includes 22 undergraduate ethics courses and 10 graduate level ethics courses (MIT 2016b). This is a very high quality tool that educators can use to enhance existing curricula or build new curricula. Learners can use this resource on their own as well.

National Ethics Center for Professional and Research Ethics hosts and maintains Ethics CORE (collaborative online resource environment). This site has a number of educational tools ranging from shared research and networking to research literature search features. The most useful feature for ethics education is the "Teaching Materials" page. This page includes active learning exercises (i.e. role-playing and mini-challenges), introductory topic packets, educational videos, course lectures, sample syllabi, and instructor resources. In addition to these, there are also "self-contained learning modules and tutorial" (i.e. reusable learning objects, RLOs). These learning modules or RLOs range in topic from community engagement and informed consent to engineering and research ethics (Ethics CORE 2016). These resources are available free of charge.

The Ethics Lab at the Kennedy Institute of Ethics (Georgetown University) is an innovate approach to addressing real-world problems by using design as strategy in conjunction with a team-based approach. There are several on-going projects supported by the lab that encourage external collaboration and creativity. The Ethics Lab also provides resources for Georgetown students including high impact education opportunities. One such opportunity surrounds the creation of "studio courses" in which students are organized in collaborative teams and conduct "intensive project-oriented research (Kennedy Institute of Ethics 2016)." The Ethics Lab is a new initiative and one component that is still in the development phase is the project incubation component. This arm of the initiative is meant to facilitate collaborate partnerships. This effort and those similar to it hold a lot of potential to bring collaborative groups together and creating meaningful social change. Looking ahead, this could be an excellent avenue to generate global ethics education partnerships, tools, and collaborate across disciplines.

The UNESCO initiative Learning to Live Together included an Ethics Education Programme as a priory in 2002–2007 (ten Have 2006). A number of activities were undertaken to address global ethics education, including: mapping of experts, sampling of teaching programmes, developing a core course in bioethics, ethics teacher training, education resources, identifying UNESCO Chairs, cultivating regional centers, promoting international networking, and providing technical assistance (UNESCO 2016a, b). There are a few helpful pieces of information on the education resources page including sample syllabi, and two casebooks. However, much of this information has not been recently updated, and there have not been any reports as to whether the Programme goals and outcomes were met. However, several new UNESCO initiatives have started including Education 2030, which is a global education initiative that has potential to facilitate ethics education and discourse across cultures and geographic regions (UNESCO et al. 2016).

7.4 Bringing Theory to Practice

One very important caveat about most of the resources addressed in the previous sections is that they all in the English language, and thus share a conceptual system (i.e. a western perspective). Therefore, it is up to the educator to use these tools and adapt them to have a global focus and reflection on the appropriate global context and nuances. The following section will provide some pedagogical insights for how this might be accomplished.

7.4.1 Addressing Content and Pedagogy

Having an abundance of resources is not always helpful if there is not a coherent way to organize these raw resources and create a meaningful learning experience. In the section on Pedagogical Resources a number of resources included more robust and well-organized teaching tools, however, for the other sections all that exist is raw content. This section will examine how to use this raw content to produce a coherent and meaningful education tools. Education alone cannot solve the problem of unethical conduct. There is no magic or silver bullet. However, meaningful and integrative education can produce learners that are, at the very least, prepared. This section will address three key elements that will assist educators in using the raw content described in the previous sections. These elements include learning outcome development, learning assessment, and experiential learning.

7.4.2 Learning Outcome Development

This chapter does not aim to detail what the most appropriate learning outcomes are for global ethics education. Arguably based on the cultural and geo-political context of a given area there will be nuanced difference between regions. Yet, pedagogy cannot be adequately addressed without at least giving a word to learning outcomes. The first, and arguably most challenging, aspect of global ethics education is identifying the learning outcomes. That is to say—after a learner completes a learning activity, experience, assignment, or course—what content knowledge, disposition, or skillset they have mastered and be able to carry to the future. This identification of the learning outcome is critical because it will drive the content, pedagogy, and assessment (Kuh et al. 2014).

There are three distinct categories of learning outcomes, they include content knowledge, attitudinal or dispositional, and skills. Content related outcomes address the information learners must know and understand, for example, the interconnectedness and interdependency of global systems and cultural competencies. Dispositional learning outcomes address those modes of being educators aim to instill in learners, for example reduce prejudice, improve cultural self-awareness, and open-mindedness. Skills related outcomes include learning outcomes geared toward developing a specific capacity within the learner, for example communication skills and ethical reasoning skills. It is up to the expert educator to determine the outcome and then strategize through evidence-based pedagogy and assessment techniques how the student can be given the opportunity to achieve the outcome (Kuh et al. 2014; Suskie and Banta 2009).

There are sets of recognized intellectual and practical skillsets relating to ethics that have emerged over the past decade. Different organizations have characterized and synthesized these skillsets in a number of different ways, yet common threads prevail. Key skillsets include: “Inquiry and analysis; Critical and creative thinking;

Written and oral communication; Information literacy Teamwork and problem solving; Civic knowledge and engagement—local and global; Intercultural knowledge and competence; Ethical reasoning and action; and, Foundational and skills for lifelong learning" (Hovland 2009). This could potentially be a starting point for thinking about global ethics learning outcomes.

7.4.3 Learning Assessment

When appropriate learning outcomes are developed the next great challenge is meeting and achieving those outcomes. The obvious question is, "how do you know when the learner has achieved the learning outcome?" Learning assessment is essentially making a judgment of whether the learner has mastered the content or skill to desired extent. Special attention must be given to assessment, as it is the educator's role to provide opportunities for students to demonstrate their learning (Banta and Palomba 2014). This may demand multiple modalities of assessment, as not all learners are able to express their learning in the exact same way (Hughes 2009). There are difficulties in assessing ethics related outcomes. This is due to the fact that there is often a dispositional component to them. However, just because it is a challenge does not mean it should be avoided, to do so could produce some undesirable outcomes (Matchett 2008; Knight and Page 2007).

In the last decade there have been a number of attempts to help structure educator thinking around assessment of ethics-related outcomes. One of the most dominant methods in the United States has been the use of assessment rubrics. This was spurred on the AAC&U's VALUE (Valid Assessment of Learning in Undergraduate Education) Rubric Initiative (Association of American Colleges and Universities 2007a). This initiative began in 2007 and involved teams of faculty and educational professionals from more than 100 higher education institutions. Over 2 years, 16 VALUE Rubrics were developed and tested by the faculty with their students at more than 100 institutions. Between 2010 and 2014, the VALUE rubrics have been viewed by more than 3300 colleges and universities world-wide (Association of American Colleges and Universities 2007b). These are widely regarded as a best-practice resource in student learning assessment as they establish baseline criteria for learner proficiency and mastery of a given outcome. The primary VALUE Rubrics appropriate for global ethics education assessment include: (1) Civic Engagement—local and global; (2) Intercultural Knowledge and Competence; (3) Ethical Reasoning, and (4) Global Learning. These assessment tools can be used as a starting point and adapted to the cultural context of institution utilizing them (Association of American Colleges and Universities 2016). VALUE Rubrics are available free of charge on the AAC&U website.

7.4.4 *Ethics Pedagogy*

The literature clearly reflects that the pedagogical style used has a direct connection to student learning and in this case the learner's level of ethical growth, understanding, and reasoning (Jurkiewicz 2002; Banta and Palomba 2014). It is imperative to note that the learning outcome and method of assessment should drive the pedagogical approach used to communicate the content. This section will provide educators with a few recognized best practices in pedagogical approaches in ethics education, global education, and global ethics education, including experiential learning and high-impact practices, as well as few unique tools and resources.

7.4.4.1 Experiential Learning and High Impact Practices

Meeting learning outcomes can be a daunting task, especially when it comes to dispositional outcomes embedded in ethics education. This is because they are more difficult to assess. It is relatively easy to determine whether a learner can apply a theory or principle appropriately and arrive at a well-justified ethical decision for a course of action, however, ascertaining whether that learner will actually make the ethical decision in a real-world context is a very different question. At this point the best we can hope to achieve is to demonstrate to the learner that making the ethical decision is not only the "right" thing to do, but also the advantageous thing to do.

In 2007, the Association of American Colleges & Universities (AAC&U) asked employers to reflect and evaluate on recent college graduates and their career preparedness in 12 specific areas. The category receiving the lowest score was "Global Knowledge", as only 18% of employers rated recent college graduates as well prepared. Forty-six percent of employers surveyed responded that graduates were not well prepared. The truly telling statistic, however, was that employers remarked that global knowledge was lacking to the extent that it would limit advancement (Association of American Colleges and Universities 2007). In 2013, the AAC&U commissioned a report that was carried out by HART Associates that surveyed employer priorities and consensus on college learning outcomes. It was found that 78% of employers strongly agree that global issues and knowledge about societies and cultures outside the U.S. is critical. Eighty-seven percent strongly agreed that knowledge of ethical issues and public debates was critical to their field. Sixty-four percent of employers indicated that colleges and universities should to place greater emphasis on ethical decision-making skills (Hart Reserach Associates 2013). When it comes to ethical conduct in the professional setting recent reports raise some alarming statistics. KPMG regularly conducts and publishes a report titled *Integrity Survey*. In their most recent report (2013) the overall reported levels of misconduct remained essentially the same from the 2009 and 2005 reports. Seventy-three percent of employees surveyed reported that they witnessed misconduct in the previous

12-months. Additionally, of the 73% of employees who reported misconduct, 56% reported that what they witness could cause "a significant loss of public trust if discovered" (KPMG 2013). These reports capture a rather disrobing trend. Learners are not prepared to deal with the ethical challenges presented in their careers and life, and we see this playing out in very real ways.

It has been demonstrated through numerous reports and publications that ethics education is best accomplished through active and experiential learning as opposed to passive learning. Passive learning does not engage students directly and takes the traditional form of the "sage on the stage", where an educator lectures to the learners and assigns texts for reading in hope of filling the learner up with knowledge. Active learning on the other hand engages the student in the learning process, and the educator plays a key role in facilitating this process (McCormick et al. 2011; Moore 2008). This engagement can take a number of forms including, but not limited to, asking students demonstrate understanding or apply knowledge through discussion, think-pair-share exercises, case studies, games, activities, or utilizing the Socratic method. Experiential learning takes active learning to the next level. Experiential learning is characterized as "a deeper understanding of subject matter than is possible through classroom study alone; the capacity for critical thinking and application of knowledge in complex or ambiguous situations; the ability to engage in lifelong learning, including learning in the workplace" (Eyler 2009, p. 26). When thinking of this definition in relation to ethics education it is clear that deeper learning occurs when students engage the material. For example, if an educator simply explains the principle of autonomy to learners through lecture it may not be a very rich experience for the learner. Whereas facilitating a discussion about the principle in relation to a case study would enhance engagement and produce a deeper learning experience. What does all of this have to do with the resources contained in this chapter, one may wonder? The answer is: everything. A video, an article, a book, or a reusable learning object, are all passive modalities unless the educator facilitates an active learning environment. Fortunately, there are resources available to help educators think through these pedagogical challenges.

One of the best compilations of experiential learning activities pertaining to ethics and social responsibility outcomes is titled *Promising Practices for Personal and Social Responsibility: Findings from a National Research Collaborative* (O'Neill 2012). This report provides data on how teaching is conducted, the value-added for learners, and how it can be improved. O'Neill specifically addresses the areas of diversity and perspective taking practices, service learning and volunteering, as well as a host of other recommendations educators and institutions can adapt to enhance student learning in ethics and social responsibility. In addition to O'Neill's publication many others exist that detail specific activities and assignments designed to motivate and engage learners, including the following:

- *Developing and Assessing Personal and Social Responsibility* (Reason 2013)
- *Engaged Learning: Enabling Self-Authorship and Effective Practice* (Hodge et al. 2009)

- *Race-Conscious Student Engagement Practices and the Equitable Distribution of Enriching Education Experiences* (Harper 2009)

It is absolutely crucial to note that these resources do not provide educators with a "here is what you should do" approach. Rather each of these presents a series of techniques, examples, and case studies that educators may adapt to their specific education context.

High impact practices (HIPs) are an evidence-based pedagogical approach very much inline with experiential learning. HIPs are teaching and learning practices that have been proved to be beneficial to learners from very diverse backgrounds. HIPs can take several different forms depending upon the learner, learning context, institutional context, and institutional priorities. HIPs at their very core are active and engage students in the learning process. The Association for American Colleges and Universities (AAC&U) recognizes ten distinct HIPs, including: first-year seminars and experiences, common intellectual experiences, learning communities, writing-intensive courses, collaborate assignments and projects, undergraduate research, diversity and global learning, service or community-based learning, internships, and capstone courses (Kuh 2008).

There is an existing and growing amount of evidence that HIPs not only facilitate deeper learning but students recognize the added value of these practices to their learning. The use of HIPs during the first year of education has been linked with higher retention and graduation rates (Bonet and Walters 2016). Not only have HIPs been demonstrated as effective among traditional college students but also among transfer students and adult learners. One student demographic of particular interest and focus, especially in context of global ethics, are students from traditionally underserved communities or regions. A report was commissioned in 2013, by the AAC&U, to assess underserved students' engagement in HIPs. This report demonstrated that there were significant academic gains, not only in learning but also in students' perception of their learning (Finley and McNair 2013). For example, among first generation college students, students reported an average boost in deep learning experiences of 11% if they participated in 1–2 HIPs, 24% if they participated in 3–4 HIPs, and 35% if they participated in 5–6 HIPs (Finley and McNair 2013). There are very similar trends among minority student populations as well as transfer student populations.

Ethics education organically fuses with active-learning and experiential pedagogy. Lecturing to learners about another culture's ethical practices does not carry the same education impact as actually having that student engage in that culture—even if it is only in a limited way. It is important to note that much of global ethics education will not happen in a traditional four-year intercollegiate setting. However, that does not mean that the same pedagogical approaches will not be highly beneficial in other contexts. Unique challenges will arise that require educators to adapt existing best practices.

7.4.4.2 Other Pedagogical Tools and Resources

There are many concrete pedagogical tools and examples in ethics education as well as sites that collect these examples. Several of these are addressed below in the Carnegie Council holding, the Ethics Education Library holdings, Intergroup Dialogue pedagogy, reusable learning objects, and ethics across the curriculum initiatives.

The Carnegie Council for Ethics in International Affairs provides a range of resources specifically geared toward the teaching of ethics content. This includes listings of films that address ethical dimensions of international affairs, historical events, policies, and human conditions (Carnegie Council for Ethics in International Affairs 2015). Sample syllabi, worksheets, lesson plans and recommended readings are also included on their website with no required fee or subscription (Carnegie Council for Ethics in International Affairs 2016a, b).

The Ethics Education Library was developed and is maintained by the Illinois Institute of Technology. This resource was developed as a result of National Science Foundation funding. This site contains a collection of case studies surrounding traditional ethics themes. Pedagogical methods and materials that were developed by scholars around the world are available, along with sample syllabi and assessment methods for measuring the effectiveness of the education and student growth (Center for the Study of Ethics in the Professions 2016).

The University of Michigan has developed a nationally recognized program in the US called the Program on Intergroup Relations. This program assists learners in developing communication skills, educates them about social justice issues and prepares them to become leaders to develop a more inclusive campus and world (Michigan 2016). This pedagogical approach promotes communication and serious discourse around controversial issues. With the primary aim of addressing diversity and social justice, this pedagogical approach leads itself very well to the challenges presented to global ethics education.

Reusable learning objects (RLOs) are independent units that can standalone and be used it multiple learning contexts. RLOs are a mechanism for granulating curricula into bite-size chunks that can be reorganized, reused and adapted to different contexts (Bath-Hextall, Wharrad and Leonardi-Bee 2011; Kong et al. 2007; Polsani 2003). For example, if there is an RLO on the function of the principle of autonomy in relation to informed consent, any educator teaching on this topic could utilize the RLO. RLOs are content specific, not course specific. The example above could be utilized in a nursing, philosophical ethics, or a science, technology, engineering and mathematics (STEM) research methods course.

The University of Nottingham's team for Health E-Learning and Media have produced a number of RLOs and made them available to the public free of charge. A range of disciplines is available, and there are a number of RLOs that address issues in medical ethics specifically. For example, topics such as moral theories, informed consent, and confidentiality are among the most common (University of Nottingham 2016). These are convenient tools has they can be immediately integrated into a given curriculum. Many colleges, universities and for-profit organization do not make their RLO repositories available online. As higher education moves toward competency-based programs the use of RLOs will continue to rise, but it is likely that access will remain very limited. Using the videos, podcasts, articles and other resources mentioned above it is possible to develop new RLOs.

Ethics Across the Curriculum initiatives have become an emerging trend in U.S. higher education. It is an effort to stimulate scholarship on ethics as well as enhance the teaching of ethics in all areas of the academy. The goal of the initiative is to transition out of the "once and done" attitude of a singular ethics course and shift to a distributed competency where students are exposed to ethical concepts, instruction and skillsets at multiple point in the curriculum. These initiates are especially useful in professional programs and general education programs (Matchett 2008). In addition to ethics being embedded in the culture of an institution, this type of work also provides space for faculty-faculty, faculty-student, and faculty-external partner collaborating and research sharing. One of the biggest resources for the initiative is the *Teaching Ethics* journal published by the Society for Ethics Across the Curriculum at the Center for Ethics, Utah Valley University (Society for Ethics Across the Curriculum 2016). A number of universities have undertaken these curricular-wide initiatives including Clemson University, Utah Valley University, Carlow University and Illinois Institute of Technology.

7.4.5 Global Ethics Centers and Organizations

Non-profit and academic centers provide some of the richest forums for global ethics discourse and educational opportunities. This is not an exhaustive list of all ethics-related centers. The list prioritizes those centers that have a global focus as opposed to centers that have a regional focus or an area of specialization (i.e. business ethics, biomedical ethics, etc.). This list can be used as a springboard to investigate the opportunities for learning and collaboration.

Selected Centers in the United States

- Center for Global Ethics and Politics located at The Graduate College and Hunter College, City University of New York, NY
- Carnegie Council for Ethics in International Affairs, The Council is located in New York, NY
- The Institute for Global Leadership located at Tufts University in Medford, MA
- Kirwan Institute for the Study of Race and Ethnicity is located at The Ohio State University in Columbus, OH
- Notre Dame Deloitte Center for Ethical Leadership located in Notre Dame, IN
- Institute for Ethical Leadership located at Rutgers Business School in Newark, NJ
- The Kenan Institute for Ethics at Duke University located in Durham, NC
- Texas Tech University (TTU) Ethics Center located in Lubbock, TX

- Olsson Center for Applied Ethics located at the University of Virginia's Darden School of Business in Charlottesville, VA
- Vann Center for Ethics located at Davidson College in Davidson, NC
- McCoy Center for Ethics in Society located at Stanford University in Stanford, CA
- The Inamori Center for Ethics and Excellence located at Case Western Reserve University in Cleveland, OH
- Center for Public Leadership located at Harvard University's Kennedy School of Government in Cambridge, MA
- Center for Ethics and Policy located at Carnegie-Mellon University in Pittsburgh, PA
- The Center for Professional and Personal Ethics located at Central Michigan University in Mt Pleasant, MI
- Markkula Center for Applied Ethics located at Santa Clara University in Santa Clara, CA
- Emory University Center for Ethics located at Emory University in Atlanta, GA
- Center For Ethics and Social Responsibility located at the University of Alabama in Tuscaloosa, AL
- Center for Global Ethics located at George Mason University in Fairfax, VA
- Institute for Global Ethics located in Washington, DC
- International Center for Ethics, Justice and Public Life located at Brandeis University in Waltham, MA
- The Center for Ethics and Social Justice at Loyola University in Chicago located in IL
- The Carr Center for Human Rights Policy at the Kennedy School of Government, Harvard University located in Cambridge, MA
- The Yale Center for the Study of Globalization at Yale University located in New Haven, CT

Selected International Centers

- Global Ethics Observatory, UNESCO
- DCU Institute of Ethics, Dublin City University in Dublin, Ireland
- Grotius Centre for International Legal Studies in Leiden, Netherlands
- The Hague Institute for Global Justice located in Hague, Netherlands
- The Raoul Wallenberg Institute of Human Rights and Humanitarian Law in Lund, Sweden
- Center for the Study of Globalization, University of Aberdeen, Scotland
- School of Applied Global Ethics, Leeds Metropolitan University, UK
- Center for the Study of Global Ethics University of Birmingham, UK
- Global Ethics Forum, University of Manitoba, Canada
- Wittenberg Centre for Global Ethics, Germany

7.5 Conclusion

This chapter has pulled diverse threads together for one common goal, and that is to provide educators and learners with resources and options to develop, improve, or enhance global ethics education. Print and digital resources as well as pedagogical tools and centers have been surveyed. From this, two important conclusions can be drawn. First, the current resources available specifically for global ethics education are lacking in quality and quantity. However, important strides are being made on this front including this publication, the International Journal for Ethics Education, and others like these. Second, meaningful global ethics education cannot be accomplished through the production of quality resources alone. Resources must be coupled with appropriate evidence-based pedagogy. That is to say, it is not enough to have high quality information available; we must also have an impactful way to help learners engage the content. Meaningful leaning experiences are the result of high-quality content and evidence-based pedagogy, both must be present for effective global ethics education.

References

Adams, A.A., S.A. Williams, and T.R. Liyanagunawardena. 2013. MOOCs: A systematic study of the published literature 2008–2012. *The International Review of Research in Open and Distributed Learning* 14 (3): 202–227.

Association of American Colleges and Universities. 2007a. *College Learning for the New Global Century: A Report from the National Leadership Council for Liberal Education and America's Promise.* Washington, DC: Association of American Colleges and Universities.

———. 2007b. *VALUE Rubric Development Project.* Association of American Colleges and Universities. https://www.aacu.org/value/rubrics. Accessed 15 Oct 2016.

———. 2016. *VALUE Rubrics.* Association of American Colleges and Universities. https://www.aacu.org/value-rubrics. Accessed 15 Oct 2016.

Banta, T., and C. Palomba. 2014. *Assessment Essentials: Planning, Implementing, and Improving Assessment in Higher Education.* San Francisco: Jossey-Bass.

Bath-Hextall, F., H. Wharrad, and J. Leonardi-Bee. 2011. Teaching tools in evidence based practice: Evaluation of resulable learning objects (RLOs) for learning about meta-analysis. *BCM Medical Education* 11: 18–28.

Blumenfeld-Jones, D. 2016. *Ethics, Aesthetics, and Education.* London: Palgrave Macmillan.

Bonet, G., and B. Walters. 2016. High impact practices: Student retention and engagement. *The College Student* (CUNY Academic Works) 15(2): 224–235.

Carnegie Council for Ethics in International Affairs. 2015. *Ethics on Film.* Carnegie Council for Ethics in International Affairs. http://www.carnegiecouncil.org/education/002/film/index.html. Accessed 15 Oct 2016.

———. 2016a. *Global Ethics Corner.* Carnegie Council for Ethics in International Affairs. https://www.youtube.com/user/carnegienetwork/featured. Accessed 15 Oct 2016.

———. 2016b. *Lesson Plan Ideas.* Carnegie Council for Ethics in Inernational Affairs. http://www.carnegiecouncil.org/education/002/index.html. Accessed 15 Oct 2016.

Center for Global Ethics and Politics. 2016. Center for Global Ethics and Politics. https://www.youtube.com/user/CGEPGraduateCenter/featured. Accessed 15 Oct 2016.

Center for the Study of Ethics in the Professions. 2016. *Ethics Education Library.* Illlinois Institue of Technology. http://ethics.iit.edu/eelibrary/. Accessed 15 Oct 2016.

Choi, C.J., J.B. Kim, and J. Beom. 2010. Globalizing business ethics research and the ethical need to include the bottom-of-the-Pyramid countries: Redefining the global triad as business systems and institutions. *Journal of Business Ethics* 92 (Supp. 2): 299–306.

Coursera. 2016. *Courses.* Coursera. https://about.coursera.org/. Accessed 15 Oct 2016.

Daniel, J. 2012. Making sense of MOOCs: Musings in a maze of myth, paradox and possibility. *Journal of Interactive Media in Education* (3): 1–20.

Documentary Heaven. 2016. *Documentary Heaven.* http://documentaryheaven.com/. Accessed 15 Oct 2016.

Ethics CORE. 2016. *Teaching Materials.* National Center for Professional and Research Ethics. https://www.nationalethicscenter.org/content/article/82. Accessed 15 Oct 2016.

Ethics for Everyone. 2016. *Ethics Movies.* http://ethicsonlineireland.weebly.com/ethics-movies.html. Accessed 15 Oct 2016.

ETHICSWEB. 2016. Science in Society Programme. http://www.ethicsweb.eu/node/1. Accessed 15 Oct 2016.

Eyler, J. 2009. The power of experiential education. *Liberal Education* 95 (4): 24–31.

Finley, A., and T. McNair. 2013. *Assessing Underserved Students Engagment in High-Impact Practies.* Washington, DC: The Center for Urban Education, University of Southern California, Association of American Colleges and Universities.

Georgetown University. 2016a. *Bibliographies on Bioethics Topics.* Georgetown University. https://bioethics.georgetown.edu/library-materials/bioethics-research-library-databases/ethx-web/bibliographies-bioethics-topics/. Accessed 15 Oct 2016.

———. 2016b. *EthxWeb: Literature in Bioethics.* Georgetown University. https://bioethics.georgetown.edu/library-materials/bioethics-research-library-databases/ethxweb/. Accessed 15 Oct 2016.

Global Oneness Project. 2016. *Global Oneness Project.* Kalliopeia Foundation. https://www.globalonenessproject.org/. Accessed 15 Oct 2016.

Globethics. 2016. *Global Digital Library on Ethics.* http://www.globethics.net/library/libraries-home. Accessed 15 Oct 2016.

Godsey, M. 2016. The value of using podcasts in class. *The Atlantic.* March 17. http://www.theatlantic.com/education/archive/2016/03/the-benefits-of-podcasts-in-class/473925/. Accessed 15 Oct 2016.

Godwyn, M. 2015. *Ethics and Diversity in Business Management Education.* Berlin/Heidelberg: Springer-Verlag.

Harper, S. 2009. Race-conscious student engagement practices and the equitable distribution of enriching educational experiences. *Liberal Education* 95 (4): 38–45.

Hart Reserach Associates. 2013. *It Takes More than a Major: Employer Priorities for College Learning and Student Success.* Washington, DC: Association of American Colleges and Universities.

Hodge, D.B., M. Marcia, and C. Haynes. 2009. Engaged learning: Enabling self-authorship and effective practice. *Liberal Education* 95 (4): 16–23.

Hovland, K. 2009. Global learning: What is it? Who is responsible for it? *Peer Review* 11 (4): 4–7.

Hughes, C. 2009. *The Assessment of Ethics.* 4th Asia Pacific Conference on Educational Integrity (4APCEI). University of Wollongong, NSW: 4APCEI, September 28–30: 1–12.

Ilkilic, I. 2014. "Culture and ethical aspects of truth-telling in a value pluralistic society" in Global Bioethics and Human Rights: Contemporary Issues. Pgs. 185–200. Editors W. Teays, J.S. Gordon, and A.D. Renteln. Maryland: Rowman and Littlefield.

Index, The Philosopher's. 2016. *The Philosopher's Index.* Philosopher's Information Center. http://philindex.org/. Accessed 15 Oct 2016.

Jurkiewicz, C. 2002.The influence of pedagogical style on students' level of ethical reasoning. *Journal of Public Affairs Education* (NASPAA) 8(4): 263–274.

Kennedy Institute of Ethics. 2016. *Ethics Lab.* Georgetown University. https://ethicslab.georgetown.edu/. Accessed 15 Oct 2016.

Knight, P., and A. Page. 2007. *The Assessment of 'wicked' Competences: A Report to the Practice-Based Professional Learning Centre for Excellence in Teaching and Learning in the Open University.* Open University. http://www.open.ac.uk/opencetl/sites/www.open.ac.uk.opencetl/files/files/ecms/web-content/knight-and-page-(2007)-The-assessment-of-wicked-competences.pdf. Accessed 15 Oct 2016.

Kong, H. P., V. Y. Lak, and K. P. Lee. 2007. *Developing Resuable Objects for Assessment.* Proceedingof the 7th WSEAS international conference of distance learning and web engineering. Beijing, China.

KPMG. 2013. *KPMG Forensic: Integrity Survey 2013*. Delaware: KPMG.

Kuh, G.D. 2008. *High-Impact Educational Practices: What They Are, Who Has Access to Them, and Why They Matter*. Washington, DC: Association of American Colleges and Universities.

Kuh, G., N. Jankowski, S. Ikenberrt, and J. Kinzie. 2014. *Knowing What Students Know and Can Do The Current State of Student Learning Outcomes Assessment in U.S. Colleges and Universities*. Urbana: National Institute for Learning Outcomes Assessment, National Institute for Learning Outcomes Assessment.

Library of Congress. 2016. *Films, Videos.* Library of Congress. https://www.loc.gov/film-and-videos/. Accessed 15 Oct 2016.

Matchett, N.J. 2008. Ethics across the curriculum. *New Dimensions in Higher Education* 142: 25–38.

McCormick, A., J. Kinzie and A. Korkmaz. 2011. *Understanding evidence-based improvement in higher education: The case of student engagement*. Annual Meeting of the American Educational Research Association. New Orleans: Indiana University Center for Postsecondary Research.

Michigan, University of. 2016. *About The Program on Intergroup Relations*. https://igr.umich.edu/about. Accessed 15 Oct 2016.

MIT. 2016a. *About OCW.* MIT. https://ocw.mit.edu/about/. Accessed 15 Oct 2016.

———. 2016b. *Course Finder.* MIT. https://ocw.mit.edu/courses/find-by-topic/#cat=humanities&subcat=philosophy&spec=ethics. Accessed 15 Oct 2016.

Moore, S.L. 2008, 142. Practical approaches to ethics for colleges and universities. *New Directions for Higher Education*: 1–7.

Nash, R., and J. Jang. 2016. *Teaching College Students How to Solve Real-Life Moral Dilemmas: An Ethical Compass for Quarterlifers (Critical Education and Ethics).* New York: Peter Lang Publishing.

O'Neill, N. 2012. *Promising Practices for Personal and Social Responsibility: Findings from a National Research Collaborative*. Washington, DC: Association of American Colleges and Universities.

Omonzejele, P.F. 2014. Understanding the concept of vulnerability from a Western Africa perspective. In *Global Bioethics and Human Rights: Contemporary Issues*, ed. W. Teays, J.S. Gordon, and A.D. Renteln, 261–272. Maryland: Rowman and Littlefield.

PhilPaper. 2016. *Welcome to PhilPapers*. Center for Digital Philosophy. http://philpapers.org/. Accessed 15 Oct 2016.

Polsani, P. 2003. Use and abuse of reusable learning objects. *Journal of Digital Information* 3 (4).

Project MUSE. 2016. *Project MUSE.* The Johns Hopkins University Press. http://muse.jhu.edu/. Accessed 15 Oct 2016.

ProQuest. 2016. *ProQuest.* http://www.proquest.com/. Accessed 15 Oct 2016.

Reason, R. 2013. Developing and assessing personal and social responsibility in college. *New Directions for Higher Education* 164: 13–22.

Rossouw, G.J. 2011. A global comparative Analysis of the global survey of business ethics. *Journal of Business Ethics* 104 (Supp 1): 93–101.

Schnackenberg, H., and B. Burnell. 2015. *The Ethics of Cultural Competence in Higher Education.* Boca Raton: CRC Press.

Society for Ethics Across the Curriculum. 2016. *Teaching Ethics*. Utah Valley University. http://www.rit.edu/~w-ethics/seac/Teaching%20Ethics.html. Accessed 15 Oct 2016.

Suskie, L., and T. Banta. 2009. *Assessing Student Learning: A Common Sense Guide*. San Francisco: Jossey-Bass.

TED. 2016. *TED Ideas Worth Spreading*. http://www.ted.com/talks?sort=newest&topics%5B%5D=Global+issues. Accessed 15 Oct 2016.

Ten Have, H. 2006. The activties of UNESCO in the area of ethics. *Kennedy Institute of Ethics Journal* 16(4): 333–351.

———. 2015. Globalization of bioethics education. In *Bioethics Education in a Global Perspective: Challenges in Global Bioethics*, ed. H. ten Have, 1–19. Dordrecht: Springer.

The Kenan Institute for Ethics. 2016. *Ethics Film Series*. Duke University. http://kenan.ethics.duke.edu/events/ethics-film-series/. Accessed 15 Oct 2016.

U.S. National Library of Medicine. 2016. *PubMed*. U.S. National Library of Medicine. https://www.ncbi.nlm.nih.gov/pubmed. Accessed 15 Oct 2016.

UMN Center for Bioethics and UMN Computer Science Department. 2016. *EthicShare*. University of Minnesota. https://www.ethicshare.org/. Accessed 15 Oct 2016.

UNESCO. 2016a. *Global Ethics Observatory (GEObs)*. Paris: UNESCO. http://www.unesco.org/new/en/social-and-human-sciences/themes/global-ethics-observatory/. Accessed 15 Oct 2016.

———. 2016b. *Learning to Live Together Activities*. Paris: UNESCO. http://www.unesco.org/new/en/social-and-human-sciences/themes/bioethics/ethics-education-programme/activities/. Accessed 15 Oct 2016.

UNESCO, et al. 2016. *Education 2030: Incheon Declaration and Framework for Action for the Sustainable Development Goal 4*. UNESCO: Incheon.

University of Miami Ethics Program. 2016. *Ethics Film Series*. University of Miami. http://www6.miami.edu/alumni/education/ethics.htm. Accessed 15 Oct 2016.

University of Nottingham. 2016. *Health E-Learning and Media Team RLOs*. School for Health Sciences. http://www.nottingham.ac.uk/nursing/sonet/rlos/rlo_tag.php?tagnum=147. Accessed 15 Oct 2016.

Wartenberg, T., and J. Akeret. 2016. *What's the Big Idea?* http://whatsthebigideaprogram.com/. Accessed 15 Oct 2016.

Weaver, G.R. 2001. Ethics programs in global businesses: Culture's role in managing ethics. *Journal of Business Ethics* 30 (1): 3–15.

Wee, C. 2014. Confucianism and killing versus letting die. In *Global Bioethics and Human Rights: Contemporary Issues*, ed. W. Teays, J.S. Gordon, and A.D. Renteln, 248–260. Maryland: Rowman and Littlefield.

Part III
Practices of Global Ethics Education

Chapter 8
You Can't Go Home Again – Bioethical Reflections on War Veterans and Refugees as Survivors; Its Implications for Global Bioethics Education

Jan Helge Solbakk

Abstract "In war, truth is the first casualty". Allegedly, it was the ancient Greek playwright, Aeschylus, who first coined this phrase. Aeshylus, the father of this art form, together with his two contemporary colleagues, Sophocles and Euripedes, were also among the first poets to put in verse and dramatize the fate of women and children as victims of this form of man-made disaster. The citizens of the ancient Greek society admired their war-plays and it was to the theatre they went when they wanted to watch moral conflicts displayed in vivo. It is hardly an exaggeration to say that in that time period of human history the theatre was considered the 'locus classicus' of ethics education. In this essay it is argued that this ancient art form still has a role to play when it comes to ethics education. Furthermore, it is argued that the moral conflicts staged in the war plays of the ancient Greek playwrights is more relevant than ever, and that teaching about the human upheavels and moral injuries caused by war, unrest and armed conflicts therefore should be included in the curriculum and acknowledged as one of the biggest challenges global bioethics education is faced with today. Three of the tragedies of Sophocles will be made use of: Ajax, Antigone and Oedipus the King, in combination with a contemporary dramatization of the human upheavals and moral injuries affecting women and children in the wake of armed conflicts: the film, *Incendies*.

8.1 Introduction

The title of this essay is inspired by Thomas Wolfe's novel with the same name, published posthumously in 1940. Allegedly, the title cropped up in a conversation Wolfe had with another writer, Ella Winter, whose remark is said to have been: "Don't you know you can't go home again?" (Godwin 2011, p. xii). The bioethical

J. H. Solbakk (✉)
University of Oslo, Oslo, Norway
e-mail: j.h.solbakk@medisin.uio.no

H. ten Have (ed.), *Global Education in Bioethics*, Advancing Global Bioethics 10,
https://doi.org/10.1007/978-3-319-78984-2_8

challenges pertaining to the fate of women and children as survivors of armed conflicts will be addressed together with the strive of refugees and internally displaced persons from different war zones in the world to come to terms with their moral injuries and losses and attempt to build a new home and restore the sense of being at home.

The aim of this essay is to illustrate how arts and humanities may be used as instruments to didactically address these global problems. The arts sources that will be made use of are three of the tragedies of the ancient Greek playwright, Sophocles: Ajax, Antigone and Oedipus the King. These plays dramatize in different ways the human upheavals and moral injuries following in the wake of war, unrest and armed conflicts. These ancient poetic sources will be used in combination with a contemporary dramatization of the same problems: The film, Incendies, which is based on a play of the same name.

A report from the UN Refugee Agency (UNHCR), *Global Trends: Forced Displacement*, indicates that in 2015 there were around 65,3 million victims of war and armed conflicts; of which 21,3 million count as refugees having had to flee their home countries, 40,8 million live as internally displaced persons and 3,2 million were asylum seekers (UNHCR 2015). This is the highest level of forced displacement ever recorded, and most alarmingly 51% of the world's refugees in 2015 were children, i.e. persons under the age of 18 years (UNHCR 2015, p. 52).

The total number of war veterans worldwide has also reached unprecedented figures. The World Veterans Federation represents around 45 million veterans worldwide (http://www.wvf-fmac.org/). The National Center for Veterans Analysis and Statistics in the USA projects that as of 2017 there are approximately 22 million war veterans only in the USA (https://www.va.gov/vetdata/Quick_Facts.asp). 8% of literally homeless persons in the USA are veterans (47,725), and if one adds to that figure the number of war veterans suffering from a sense of homelessness – of living internally displaced lives – the reach of this problem becomes even more evident (National Alliance to end Homelessness 2016, p. 7).

In a globalized world these aftermaths of man-made disaster is no longer confined to the most remote parts of this burdened planet. On the contrary, today's wars, unrest and armed conflicts, generate forms of human suffering that make distance and borders morally irrelevant. Millions of victims of war are knocking on the doors of people enjoying peace and prosperity in search of shelter and protection. The largest refugee crisis ever is a fact of this world, and of this time period of human history. In a book about global bioethics education for today these challenges should be included in the curriculum and acknowledged as one of the biggest challenges bioethics is faced with. The Preamble to the Constitution of UNESCO declares: "since wars begin in the minds of men, it is in the minds of men that the defences of peace must be constructed" (http://portal.unesco.org/en/ev.phpURL_ID=15244&URL_DO=DO_TOPIC&URL_SECTION=201.html). For a bioethics education with global aspirations this phrase can serve as a reminder of the precariousness of the human condition and as a call to include in the curriculum teaching about the human upheavals and moral injuries caused by war, unrest and armed conflicts.

8.2 The Use of Tragic Narratives – Some Didactic and Methodological Considerations

The suggestion to make use of the ancient Greek poetic sources here mentioned draws on Aristotle's differentiation in the *Poetics* of tragedy from history and his claim that tragic poetry is more *philosophical* than history since it speaks of things which at all times prove to be morally true and relevant, while history narrates contingent events or incidents (Aristotle 1984, *Poetics* 51b8–12). In addition, the approach here suggested is inspired by Bernard Williams' statement in *Shame and Necessity* about the moral timelessness of these poetic sources:

> When the ancient speaks, they do not merely tell us about themselves. They tell us about us. They do that in every case in which they can be made to speak, because they tell us who we are. That is, of course, the most general point of our attempts to make them speak. They can tell us not just who we are, but who we are not: they can denounce the falsity or the partiality or the limitations of our images of ourselves (Williams 1993, pp. 19–20).

Another reason for making use of tragic narratives relates to the concept of 'catharsis', a concept Aristotle makes use of to indicate the didactic impact a tragic play may have on its audience. Says Aristotle in his famous definition of a tragedy:

> Tragedy is a representation of a serious, complete action which has magnitude, in embellished speech, with each of its elements [used] separately in the [various] parts [of the play]; [represented] by people acting and not by narration; accomplishing by means of pity and terror the catharsis of such emotions (Aristotle 1984, *Poetics* 49b23-27).

In the secondary literature no substantial consensus has been reached as regards the exact meaning Aristotle attributes to tragic catharsis. On the contrary, a whole range of seemingly disparate interpretations have been suggested, all alluding to some sort of clarification and cleansing:

- *medical* interpretations and interpretations of catharsis as a natural process of *discharge/release* or *outlet* of emotions,
- catharsis conceived of as emotional and intellectual *clarification*,
- *moral* interpretations, including interpretations of catharsis as an *education* of the emotions,
- catharsis conceived of as the experience of emotional *relief*,
- *aesthetic* interpretations or interpretations of a *dramatic* or *structural* nature, and finally
- complex or 'holistic' interpretations of catharsis.

Elsewhere I have argued that the uncovering of the broadness and variability of meaning and applicability of the conception of tragic catharsis paves the way for a *psycho-somatic* conception of moral enlightenment and learning (Solbakk 2006, 2015). That is, as the ancient Greek plays affected the spectator's mind and soul as well as his bodily hair and humors, in our teaching of bioethics "we should look for analogies in our experience and our sense of the world" (Williams 1993, p. 19) that have the didactic potential to implicate students in their psycho-somatic *entirety*. This is the reason why a contemporary dramatization of the human upheavals and moral injuries following in the wake of war have been included. As stated by U.v. Wilamowitz-Moellendorf: "To make the ancients speak, we must feed them with our own blood" (quoted in Williams 1993, p. 19). That is, only in such a way can a situation of 'moral simultaneity' be established between our time and the ancient Greek playwrights.

In an essay on the concept of disaster in literature to be published in a book on disaster bioethics (O'Mathuna et al. 2016) I provide a detailed analysis and interpretation of the fate of war veterans, women and children in Sophocles' play *Ajax and* Euripedes' *Andromache*, *Hecuba,* and *The Trojan women* followed by a brief presentation of two contemporary projects that draw on these plays; Bryan Doerries' *Theater of war* (Doerries 2015) and *The Queens of Syria*, a Developing Artists' project, to illustrate the moral timelessness of these plays (Developing Artists 2013). In the present essay I suggest the opposite procedure; i.e. to provide a detailed analysis of the film *Incendies* while using material from the three plays of Sophocles mentioned in the introduction to recall that the emotional sufferings and moral injuries articulated on stage in his plays continue to form the core narratives of victims and survivors of today's war and armed conflicts. The reason for having selected the film *Incendies* is that it dramatizes in condensed form the many human upheavals and moral injuries women and children are subjected to in armed conflicts.

8.3 Historical and Artistic Antecedents of the Film Incendies

Incendies is a French-Canadian film (released in 2010) based on a play from 2003 by Wajdi Mouawad – a Lebanese-Canadian writer, actor and director. *Incendies* the play, translated into English with the title *Schorched* in 2005 (Mouawad 2010, English translation) is inspired by and partly based on events narrated in the memoirs of Soha Bechara, a young Lebanese political activist and resistance fighter. Born to a family of secular Greek Orthodox Christians in Beirut, Bechara sided with the fractions fighting the Israeli invasion of South Lebanon in 1982. Disguised as an aerobic instructor she managed to get access to the home of General Antoine Lahad, the leader of the South Lebanon Army (SLA) who supported the Israeli invasion. On November 17, 1988, the then 21 old Bechara fired two shots at Lahad, seriously wounding him. She was immediately arrested and without charges or trial she was interned for ten years in the Khiam Camp in solitary confinement. This was a detention centre near the Israeli border allegedly supervised and funded by Israel (Holstun 2015, p. 3). As a result of an intense public campaign both in Lebanon and Europe, in particular in France, she was released on September 3, 1998. In 2000 she published her story, *Resistante* (English title, *Resistance: My Life for Lebanon*, 2003), where she tells about her life in Lebanon before and after becoming a political activist and resistance fighter. In 2011, she published a second book about her confinement for 10 years in the "hellhole" of Khiam, (original in Arabic, *Ahlumu bi-zinzānah min karaz*, published in French in 2014 as *La fenêtre: Camp de Khiam* and in English as *I Dream of a Cell of Cherries*). This book was written together with Cosette Elias Ibrahim, a Lebanese journalist who had been detained in the same place for nine months until Israel on the 22nd of May 2000 pulled out of the country and the South Lebanon

Army forces abandoned the Khiam Camp. Allegedly, this part of her memoirs was 'inspired' by Israel's bombing of the detention centre in 2006 (Holstun 2015, p. 32):

> After Israel's retreat in 2000, Hezbollah turned Khiam Camp into a museum of the resistance. On 25 July, 2006, Israel bombed it […] This attack persuaded former internee Cosette Ibrahim to collaborate with Bechara on a more lyrical memoir […]. Disinterested in traditional prison memoirs, Ibrahim helped Bechara write, 'not to boast about certain heroic moments or to renounce days of ordinary cowardice,' but 'to accompany Soha's words, her scattered memories, to leave traces like Post-it Notes […] so as not to forget, to engrave certain lived details of Khiam Camp: details that we cannot engrave on the walls, as prisoners habitually do… the remnants of walls that no longer exist.' With Ibrahim's help, Bechara records her lived experience of internment in short chapters: the complaint of her mother, temporarily interned, that Soha launched the operation during olive-harvesting season; the names of her fellow internees, and the dates of their arrival, departure, and death; menstruation, bruising, eczema, and body odor; emotional dialogues with Kifah about Sabra and Chatila; her poems 'as mediocre as they were passionate'; and detailed recollection of resistance through crafting olive pit rosaries, knitting, and embroidery, including a chapter titled 'In the Beginning Was the Needle'.

In connection with the premier of the film *Incendies* in Geneva in January 2010 Bechara was present and she gave an interview (for this, see Joe's Garage 2010). She critized the film for not reaching up to the level of abstraction and universality that in her opinion is the case with the play of the same name:

> Wajdi was already working on a play", Bechara recalled. 'He had seen Randa's film, had read my book [*Resistance: My Life for Lebanon*] and all these ideas were a point of departure. He built his story little by little and the play came abstract – the soul of his play was the fact that it was abstract. This was its strength. I think the director [Villeneuve] tried to remain faithful to Wajdi and to the text', Bechra continued. 'He tried to put himself in Wajdi's shoes and didn't manage. Wajdi had made a play that was universal, whereas [Villeneuve] brought it closer to Lebanon with all its cliches. I'm not even sure why it had to be in Arabic'. (Joe's Garage 2010).

Bechara's praise of the play and concurrent critique of the film is interesting when reading it in the light of Aristotle's differentiation between history and tragedy. "A universal", says Aristotle, "is the sort of thing that a certain kind of person may well say or do in accordance with probability or necessity – this is what poetry aims at, although it assigns names [to people]. A particular is what Alcibiades did or what he suffered" (Aristotle 1984, *Poetics* 51b8–12). This remark about the use of *historical* names in tragedies, and thereby about the representation of events that have actually taken place, is important, because it informs us that not everything in a tragedy is made up. More important, however, is the explanation Aristotle gives for the poet's use of historical material. For tragic accounts to be trustworthy, they must be *possible*, and things which have happened, says Aristotle, are obviously possible. Consequently, by using events, names or things that have actually existed or taken place, as templates for giving shape to a tragic plot, the poet is free to invent for himself a whole that *may* have taken place (Aristotle 1984, *Poetics* 53b23–27). Thereby, out of the creative reconfiguration of the historical and particular, emerge neither *imaginary* accounts nor wild *thought experiments* but accounts that are possible and at the same time of universal moral relevance and value (Solbakk 2006).

Neither the play nor the film makes use of the names of events, persons or places mentioned in Bechara's memoirs. All names are fictitious, and the film is true to the fictitious names used in the play. It remains, therefore, unclear why Bechara deems the film to not reach up to same level of abstraction and universal moral relevance as the play. Her somewhat negative verdict stands in contrast to the many highly positive reviews from critics, including film awards, and the successful reception the film has been met with among moviegoers world wide. In the end, only by exposing oneself both to the play and the film will it be possible for each spectator to decide which of the two artistic works triggers most the emotions Aristotle associated with tragedy and the kind of catharsis he attributed to such an exposition. Didactically speaking there are reasons to counter Bechara's critique of the film's inferiority with regard to universality, and notably for the simple technical reason that the film is easily accessible on internet both in English, French and Spanish while the complete play is only accessible as a text or as read and/or acted out on stage in French or English.

8.4 Incendies the film

You and whose army?

The first scene of the film *Incendies* is shot in a rocky and sunburnt landscape. Chirping sounds of birds are heard, followed by a song of the English rock band Radiohead, *You and Whose Army?* (The band's song can be heard at Radiohead 2001a). The music adds a forlorne and abandoned dimension to the scene. The camera lense focuses on a lonely palm tree seen through the broken windows of a shabby building where a few armed men in military boots, but not dressed in uniforms, hold guard over a group of young boys queuing to have their hair shaved off. In slow motion the camera lens sweeps from one child's face to the next, before dwelling on the face and naked feet of the boy undergoing the hair cutting. His right heel has a tatoo formed as three small dots. He stares directly into the camera with eyes wide open, a serious face filled with wonder and concern. No words between the men and the children are uttered. Only the words of the song speak; enticing words about an army, about the Roman Empire, about action and friendship, and memories that easily fade away. In short, a poetic representation of the tragic 'art' of recruiting children as soldiers (Radiohead 2001b):

Come on, come on
You think you drive me crazy
Come on, come on
You and whose army?
You and your cronies
Come on, come on
Holy roman empire
Come on if you think
Come on if you think
You can take us all on
You can take us all on

You and whose army?
You and your cronies
You forget so easily
We ride tonight
We ride tonight
Ghost horses
Ghost horses
Ghost horses
We ride tonight
We ride tonight
Ghost horses
Ghost horses
Ghost horses.

A recent report from the organization *Their World* states:

> As many as 300,000 children are believed to be serving as soldiers in armed conflicts around the world – depriving them of a normal childhood and education. These boys and girls, some as young as seven, serve in government forces and armed opposition groups. They fight on front lines, participate in suicide missions and act as spies, messengers or lookouts. Girls are often forced into sexual slavery. Many are abducted or recruited by force, while others join out of desperation, believing that these armed groups offer their best chance for survival (Their World 2017).

These figures demonstrate in a dramatic way how armed conflicts easily lead to a total instrumentalization of human lives and put children – the most vulnerable of human beings – in harms way.

Didactic Task 1

This part of the film can be used to reflect on the normative question how children living in armed conflicts should be perceived. Are they always to be considered as victims – i.e. as passive recipients of the violence of adults? This perception would imply that notions such as "fragility", "damage", "deprivation" "integrity" and "trauma" would occupy centre stage in the account, and that the moral focus would be on childrens' vulnerability and, consequently, on ways of protecting them from the human rights violations and forms of capability deprivation they are prone to suffer in the wake of violent conflict.

An alternative – albeit quite controversial and reductionistic – perception that has been proposed is that children's involvement in armed conflicts is not always the result of compulsion, coercion, and brainwashing: "Few authors have shown willingness to consider the possibility that, in some situations, young people may engage with military groups as a reasoned strategy – as the most desirable option within the range of choices available. They may also enrol out of social and political concern" (Hart and Tyrer 2006, p. 9). This perception would imply that notions such as "agency", "engagement", "resilience" and "resourcefulness" would be given a more prominent place in the narrative, but at the loss of the human rights focus and capability approach advocated in the previous account.

The last will of a mother

The next scene of the film takes place in the office of Jean Lebel, a middle-aged notary about to meet with a young man and woman, Simon and Jeanne Marwan. In the first part we see Mr. Lebel leaning towards the shelves in a room packed with archives boxes, his upper body in a slumped position, his head half bowed, his eyes closed as if meditating. The song from the first scence is still heard. Mr. Lebel opens a folder from one of the boxes with the name of Nawal MARWAN. A sigh is heard before the scene shifts to his office where Jeanne and Simon have arrived. He greets Jeanne with a kiss on each cheek and pats Simon on his right shoulder. The dialogue between them reads thus:

MR. LEBEL: I'm glad to see you.

[He hangs up his coat, and the three of them walk into his office, while on the screen 'TWINS' are written in read. A sighing Lebel takes up a paper knife, opens a sealed envelope and starts reading].

MR. LEBEL: Testament of Dr. Nawal Marwan. Opening of the will to her two children.
Jeanne and Simon Marwan. In accordance with the will, regulations and rights Nawal Marwan has asked the notary, Jean Lebel, to be executor of the will. [Here Mr. Lebel interrupts the reading for a moment to explain his own role in the case].
Hmm…I must tell you here, it is the decision of your mother. I was personally against it, but she insisted. You know your mother was more than an employee for me. My late wife and I considered you almost as part of our family. So, I could not refuse.

[The reading of the will continues].

All my assets will be shared between the twins Jeanne and Simon Marawan. The money is to be shared equally between them. And my furniture according to their desire and agreement.
Burial [a sigh slips from Mr. Lebel]. To notary Jean Lebel: Bury me without a coffin, naked. And without prayers. The face turned towards the soil, face first against the world [A sigh is heard from Simon. The face of Jeanne looks tense].
Gravestone and epitaph. No stone shall be put on my grave. And my name engraved nowhere. No epitaph for those who do not keep their promises.
To Jeanne and Simon. A child is as a knife planted in the throat. One cannot remove it easily. Jeanne, notary Lebel will hand you an envelope. This envelope is for your father. Find him and hand him the envelope. Simon, the notary will give you an envelope. [Simon looks uneasy and interrupts the reading].

SIMON: I've heard enough.
MR. LEBEL: But, it is not finished.
JEANNE: Continue.
MR. LEBEL: Simon, the notary will hand you an envelope. This envelope is for your brother. Find him and give him the envelope. When these envelopes have

been given to the addressees, a letter will be given to you. The silence is broken. A promise kept. A stone may be placed on my tomb, and my name engraved on the stone in the sun [Mr. Lebel folds together the testament, looks at the twins and sighs].

I admit it is a bit special. [Jeanne looks upset and concerned. Simon looks annoyed].

SIMON: Is there stuff to sign?

MR. LEBEL: No, not for today.

SIMON: OK. Thank you, Jean. [Simon turning towards, Jeanne]. Shall we?

JEANNE: Wait [she sighs before she turns to the lawyer]. Do you know the history of our father? You know that we have no brother. Why did you let her write this? [Again a sigh from the lawyer is heard].

MR. LEBEL: I understand your surprise. Its…very astonishing. [Simon addressed to the lawyer].

SIMON: Stop pretending the innocent. We know about what we talk. There is nothing surprising in this.

MR. LEBEL: One does not invent such things, Simon, not in a will. [Taking up a third envelope the lawyer sighs again].

There is also this [He gives the envelope to Jeanne. She opens the envelope and finds in it a passport with Arabic letters and a necklace with a Christian cross].

Well, your mother leaves us this morning with something rather….We'll sleep on it, then….We'll meet again, huh?.

SIMON: I'll wait outside. [He gets up from the chair to leave] Thank you, Jean.

MR. LEBEL: Sit down, Simon.

SIMON: Look, Jean. Your secretary loved working for you.This was good for her life, bravo! But your secretary was also my mother. Her job as mother is another story [The lawyer sighing again].

MR. LEBEL: Simon…

SIMON: Your secretary you can bury on the side you want. My mother I want to bury in the normal way. So that she for once in her crises of life does something normal. This is not negotiable [Simon leaves the office and shuts the door behind him. Jeanne remains seated. She opens her mother's passport and looks at the photo].

MR. LEBEL: Take your time. Come and see me when Simon has calmed down. I know it is very unusual, but…[he sighs] your mother was not crazy Jeanne.

The last part of this scene takes place outside the building of the notary's office. Chirping birds are heard. Simon stands in a slumped position outside a car with his hands in his pockets. Jeanne is approaching the car.

SIMON: We have a big family now [For a short while they are both standing there in silence].

She spoke of a dog? [Mockingly] A big family, there is always a dog. Will we also have to go and find a dog? It's fucking crazy.

I'll fix that with Jean.

JEANNE: Fix what?
SIMON: We will arrange a normal burial. We'll take care of her, then I'll take care of you.

They open the front doors of the car about to enter it. Janine closes the door and starts walking away from the car.

SIMON: Where are you going?
JEANNE: You bury the bin as you like.
SIMON: Oh, stop! You feel guilty because you were not there when she died. You feel guilty because you were there when she had her accident. We need to stop you feeling guilty for everything. I am not guilty of anything, myself [Jeanne, on the verge of crying].
JEANNE: Do you realize what he just read?
SIMON: Well, she's gone. It's over, the crisis. One is in peace! [Simon kicks the back of the car]. I have the fucking peace [Sighs are heard]. I'm at peace now.
JEANNE: Yes, I see that [Sighs are heard again].

Didactic Task 2

Anger and guilt are two morally charged words and emotions that play prominent roles in the film as it evolves. Simon's anger and Jeanne's alledged feeling of guilt in the previous scene are just starters. In her book, Anger and forgiveness', Nussbaum claims "that anger is always normatively problematic, whether in the personal or in the public realm" (*Nussbaum* 2016, *p. 5). In spite of this, she links the two emotions. First, by referring to the fact that guilt often is thought of as a "sort of self-anger". And second, by emphasising that both emotions focus on acts (Nussbaum* 2016, *p. 128):*

> *Guilt is a negative emotion directed at oneself on the basis of a wrongful act or acts that one thinks one has caused, or at least wished to cause. It is to be distinguished from shame, a negative reaction to oneself that has a characteristic, or trait, as a focus. We can see that guilt parallels anger: both focus on acts. This focus on acts, rather than the self, is auspicious, since separating the deed from the doer is a constructive aspect of moral (or nonmoral) change. Another promising aspect of guilt is that, focusing on an act, it typically focuses on damage to our relationships with others, an important and (unlike more or less permanent traits) a remediable aspect of our conduct. The problem comes, as usual, when we focus on the wish for payback..*

Nussbaum addresses three prevalent perceptions of anger: 1) that anger is necessary to protect one's dignity and self-respect in the face of wrongdoing; 2) that anger is a response that takes the wrongdoer seriously; and 3) that anger is an essential part of combatting justice (Nussbaum 2016, *p. 6*). She acknowledges that anger thus conceived might be "instrumentally useful", but "this limited usefulness", she asserts, "does not remove its normative inappropriateness" (Ibid.).

This part of the film may serve as a starter for discussing the relation between these morally charged emotions, and the way anger, resentment and feelings of guilt might impact the lives of women and children as survivors of armed conflicts. These emotions are richly represented in the film.

The land of solitude

The next scene starts in the university classroom where Jeanne is introduced as the new maths teacher to the class by her mentor.

THE MATHS PROFESSOR: Mathematics as you have known so far has aimed to achieve stringent and definitive answers starting from strict and definitive problems. Now you are about to enter an entirely different adventure – the subject matter of irresolvable problems that will lead you to other problem also irresolvable. People around you will repeatedly pick on you and say that what you do is useless. You will have no argument to defend yourself, because it is a matter of exhausting complexity. Welcome to pure mathematics, to the land of solitude. I present to you my assistant, Miss Jeanne Marwan (Jeanne leaves her seat in the classroom and walks towards the blackboard].

JEANNE: Hello! We will start with the conjecture of Syracuse.

This problem of arithmetics, which is also known as Collatz problem or the 3n + 1 conjecture is still unsolved, in the sense that no mathematician so far has been able to explain why the end result is always one. The problem can be stated as follows:

- If the number is even then divide it by two
- If the number is odd then multiply it by 3 and add 1

The conjecture is that this process always reaches the number 1.

Example: start with the number 10. Following the above mentioned process the sequence will be: 10 → 5 → 16 → 8 → 4 → 2 → 1.

And if the process is continued after reaching the number 1 then it indefinitely repeats the following cycle:

1 → 4 → 2 → 1…. (https://mathematicalgarden.wordpress.com/2009/01/04/the-syracuse-problem/).

As will become revealed later in the film, this conjecture contains the solution to the riddle Jeanne and Simon have been faced with in the will of their mother. The next part of this scene takes place in the office of Jeanne's professor. Evidently, Jeanne has told him about the enigmatic will of her mother.

THE MATHS PROFESSOR: What is your intuition telling you? [Jeanne sighs and looks bewildered].

Your intuition is always right. That's the reason why you have the potential of becoming a real mathematician. But here, you'll need help. Have you any family there? [Jeanne shakes her head]. Contacts?

JEANNE: It's ridiculous.

THE MATHS PROFESSOR: You have come to learn that A, your father is alive; and B, you have another brother. What is ridiculous is that you are questioning what is inevitable. You need to know. Otherwise, you will never be at peace. And no peace of mind, no pure mathematics. It provides you a starting point.

JEANNE: My father died during the war, in Daresh. That's the unknown variable in the equation.
THE MATHS PROFESSOR: It never starts with the unknown variable.
JEANNE: My mother comes from a village which is called Om, in Fouad. She studied to teach French at the University of Daresh [While they are talking, the professor brows through his diary].
THE MATHS PROFESSOR: If she has studied, there is hope. Said Hadar. You'll go to knock on the door of Said Hadar, an old friend of me who teaches at Daresh. You can trust him.

In the next part of the scene no words are spoken. An old sewing machine and a dried out pot plant are displayed, indicating the apartment of the deceased, and Jeanne in the process of looking through the drawers in the bedroom. Drawer after drawer are removed from the commode, put on the bed and searched through. Sighs are heard. At the bottom of one drawer she finds an envelope with a photo of the mother as a young woman, face stern in front of a brick wall with Arabic letters. The scene switches to the apartment of her brother, who reluctantly looks at the same photo lying on the table. Next, Jeanne is seen heavily dressed in front of a swimming pool with frozen water. A flashback sequence is following where Jeanne is swimming in the same pool with her mother sitting on a beach chair on the edge of the pool staring out as if paralysed by something. Jeanne is alarmed when her mother does not react when she addresses her. The scene shifts to a hospital. Simon is hurrying along the corridor on the search for his mother. Nawal is lying in a bed in the corridor; she does not respond to Simon's questions about what has happened and about Jeanne's whereabouts. Next, Simon and Jeanne are displayed in the sickroom with their mother. A physician is examining Nawal.

THE PHYSICIAN: Is she suffering from migraine?
JEANNE: No.
THE PHYSICIAN: Memory loss?
JEANNE: No.
THE PHYSICIAN: Temporary absence?
SIMON: It is absent in general [Jeanne with a stern look at Simon].
JEANNE: No, she was never absent.

The scene ends in silence, with Simon and Jeanne sitting in the sickroom beside their mother, who seems to be at sleep.

Love and war in the family

The next scene opens with 'NAWAL' written in read on the screen. A young woman (Nawal) is seen running in a rocky, sunburnt landscape towards a young man (Wahab).They embrace quickly, and continue to run as if trying to escape and hide away from someone. Squeal of cicadas is heard. A young man watching them from a rock just ahead of them whistles briefly and jumps down in front of them. Wahab, wearing a Palestinian keffieyh (scarf), tries to hide Nawal behind his back and they both move backwards as the other man is approaching them. In fury the young man addresses Wahab (in Arabic): "Let go my sister. Where are yo going, son of a refugee?". Another young man (probably the brother) comes running towards them with a pistol in his hand. Nawal pleads to him: "Nicolas, stop, Nicolas stop!", but he points the pistol at Wahab, shouting, "Go back home!", and shoots him. "Wahab", Nawal is screaming. Looking with contempt at the body of Wahab on the ground, Nicolas swears, "We'll seek you in your camps and slaughter you!". The other brother asks for the pistol, "Give it. I want to [kill her]. It is my duty". Nawal is on her knees with the weapon pointing to the back of her head. The brother shouts at her: "You stained our family's honor". He is about to pull the trigger when a cry from an elderly woman (Nawal's grandmother) comes to Nawal's rescue. She shouts to them: "That's enough. Come home now. Come home!".

The rest of this scene takes place in the house of the grandmother. While closing the door and all windows she is bestowing Nawal with furious charges: "What have you done! You've humiliated us, ruined the family name! Why my God, have you cast us into darkness?" [Nawal is heard sobbing. The grandmother continues the angry charges while shaking Nawal forcefully]. "My God, what have you done? Why did you do this to me? What am I to do with you! Should I kill you?" [Nawal's sobs and the complaints of the grandmother become even more audible when Nawal reveals that she is pregnant]: "I'am pregant grandma". The grandma slaps Nawal and screams: "My God, blackness everywhere! Why! All is black. Why? Why did you do this to us? Am I supposed to kill you? What am I going to do with you?"

After a while, the old and furious woman threatening to kill her own granddaughter seems to have been replaced by the comforting grandmother concerned with protecting and helping the pregnant granddaughter out of her disgrace and misery. Nawal is lying in her bed, her body crouched.

THE GRANDMOTHER: Eat. Eat if you want your child to be strong. Listen to me carefully.
There's nothing here for you now. I'll help you after the birth. I want you to leave here. You'll stay with your uncle Charbel in town. You'll go to school there. Yo'll learn to read, to think. To escape this misery. Promise me and I'll help you. Promise me you'll go to school".

NAWAL: I promise you, grandma. I promise you.

The rest of this part of the film displays Nawal's growing pregnancy, the birth of a son, the grandmother marking the right heel of the baby with a three-dotted tatoo, the same tatoo that was shown on the heel of the young boy in the queue with other boys waiting to have their hear shaved off. The grandmother hands the newly tatooed baby to Nawal: ""Take a good look at your mother. So you'll recognize her". Nawal to her son: "I'll find you again one day, I promise you my love". This is the promise Nawal refers to in her will and which she states has been broken, and that needs to be fulfilled before she can be honored with a stone and an epitaph on her grave.

In disguise of the twilight a woman is seen carrying away the baby, before the scene shifts to Nawal who is about to leave her grandmother's house. The farewell words of the grandmother echoes the frightening prophecies so often uttered in the ancient Greek tragedies: "Come, its time you left. I see terrible times ahead". Nawal repeats her promise to her son: "One day I'll find you, my son".

Didactic Task 3

Honor killing is the murder of a person accused of "bringing shame" upon their family, e.g. for refusing to enter a marriage, being in a relationship that is disapproved by their family, having sex outside marriage, dressing in ways which are deemed inappropriate, engaging in non-heterosexual relations or renouncing a faith.

In some parts of the world, women who have been raped have also been murdered for the 'dishonour' of being a victim and the 'disgrace' it brings to their family.

Honor killing is believed to have originated from tribal customs where an allegation against a woman can be enough to defile a family's reputation – 'a life without honour is not worth living.'

Honor killing is a global problem that occurs in all regions of the world.

None of the world's major religions condone honour-related crimes.

But perpetrators have sometimes tried to justify their actions on religious grounds (Source: http://www.bbc.co.uk/ethics/honourcrimes/crimesof-honour_1.shtml).

Article 12 of the Universal Declaration on Bioethics and Human Rights, 'Respect for cultural diversity and pluralism' states: The importance of cultural diversity and pluralism should be given due regard. However, such considerations are not to be invoked to infringe upon human dignity, human rights and fundamental freedoms, nor upon the principles set out in this Declaration, nor to limit their scope.

A suggestion related to this part of the film is to use the case of honor killing to discuss what respect for cultural diversity and pluralism entails, and what are the limits of such forms of respect.

Death and honor – the echoes of Ajax and Antigone in Incendies.

The society in which Nawal was raised complies clearly with the moral codes of a so-called shame culture; i.e. "a culture in which conformity of behaviour is maintained through the individual's fear of being shamed", and which "puts high emphasis on preserving honor and on not being publicly disgraced" (https://en.oxforddictionaries.com/definition/shame_culture). In such a culture honor-killing can be viewed not only as morally justifiable but as a *necessary* act to conduct in order to preserve honor, e.g. in the face of the family. For this reason Nawal's grandmother did not condemn the brothers' killing of Wahab. On the contrary, she was contemplating whether to take deadly action herself – against her granddaughter – to preserve the family honor: "My God, everything is black. Why? Why have you done this? Do I have to kill you? What will I do with you?"

Honor, shame and necessity are three morally charged words that also played important roles in the ancient Greek society. Roughly speaking, in the scholarly literature there are two opposing positions; the "progressivist" view versus a view I, with reference to Cairns (1993, p. 47), suggest to label the "similarity view". The progressivist view holds that a society emphasising the importance of shame is morally inferior to a society emphasising the role of guilt (Benedict 1946, Dodds 1951):

> The shift from a shame culture to a guilt culture [...] is taken as a sign of moral progress. Thus the warrior society represented in the Homeric epics – a shame culture [...] slowly gave way to a guilt culture, which began to emerge in fifth-century democratic Athens but did not achieve a fully developed expression in the classical world until the advent of Christianity (Konstan 2012, p. 1031).

The "similarity" view, on the other hand, challenges the progressivist hypothesis of a great shift from ancient to modern ethical thought because of the latter's emphasis on guilt (Ohly 1992, Cairns 1993, Williams 1993). Says Williams about the meaning of shame in the ancient Greek society:

> The basic experience connected with shame is that of being seen, inappropriately, by the wrong people, in the wrong condition. It is straightforwardly connected with nakedness, particularly in sexual connections. The word *aidoia,* a derivative of *aidōs*, "shame", is a standard Greek word for the genitals, and similar terms are found in other languages. The reaction is to cover oneself or to hide, and people naturally take steps to avoid the situations that call for it. (Williams 1993, p. 78)

But shame does not always come as a result of being publicly disclosed: "Even if shame and its motivations always involve in some way or other an idea of the gaze of another, it is important that for many of its operations the imagined gaze of an imagined other will do" (Williams 1993, p. 82). Williams also questions the progressivist view of a sharp distinction between the ancient Greek conception of shame and the conception of guilt, and the view that because the Greeks in the time of Homer and the ancient playwrights did not have a direct equivalent to the word "guilt", their understanding of guilt was necessarily inferior to our understanding (Williams 1993, p. 89). For this reason he argues that the ancient Greek term *aidōs* and related terms "cannot merely mean "shame", but must cover something like guilt as well" (Williams 1993, p. 90). This is a view also advocated by Cairns:

> [...] since there is no distinction to be made between shame and guilt in terms of internal versus external sanctions, and since guilt and shame can, in certain circumstances, overlap (or, if you prefer, co-occur), we should be ready to admit that a characterization of a given instance of *aidōs* in terms of shame need not rule out a characterization in terms of guilt. (Cairns, 1993, p. 4)

Preserving honor and avoiding being publicly disgraced is dramatized in Sophocles' plays *Ajax* and *Antigone* in ways that clearly echo the experiences of Nawal. The play about *Ajax*'s fall into shame starts with a conversation taking place outside his tent, during the siege of Troy, between the goddess Athena – the daughter of Zeus – and Odysseus, the legendary Greek king of Ithaca and hero of Homer's epic poem the *Odyssey*. In Sophocles' play Odysseus is depicted as the subject of Ajax's most deep-felt rancor. The topic of the conversation is the sudden madness that has taken possession of him. Odysseus has been sent by the Greek kings and war leaders, Menelaus (king of Mycenaean Sparta) and his brother Agamemnon (king of Argos), to find out whether Ajax is the vicious killer of the cattle that the Greeks had taken captive from the Trojans. Athene confirms Odysseus' suspicion and unveils to him the cause behind Ajax's madness: "He was stung by anger on account of the arms of Achilles" (*Ajax*, lines 40–41). The reason behind the rage of Ajax is that the armor of the dead Achilles, the most formidable of the Greek warriors, had been awarded to Odysseus instead of to himself. Ajax feels betrayed by Menelaus and Agamemnon.

He swears to take revenge and in the darkness of night he sets out secretly to trace and kill Odysseus as well as the brother kings. However, Athena steps in and deludes Ajax into attacking instead the cattle that the Greeks had taken from the Trojans.

When Ajax comes to his senses and realizes what he has done, he is overwhelmed by shame. His self-respect is put in peril. And his mind is dominated by the feeling of having become a complete ridicule and outcast in the world of humans and gods. As is the situation with Nawal, his fall into disgrace makes that he can't go home again – not ever:

> Shall I cross the Aegean sea, leaving behind the station of the ships and the sons of Atreus, and go home? And what kind of face shall I show to my father Telamon when I appear? How ever shall he bring himself to look at me when I appear empty handed, without the prize of victory, when he himself won a great crown of fame? The thing is not to be endured (Sophocles 1994, *Ajax*, lines 430-434, and lines 460-465).

There is for him only one viable option left, self-slaughter, an act which for the morality of a noble warrior was necessary to conduct to preserve his public honor as well his honor in the face of his father. Likewise, Nawal's fall into disgrace enforced upon here the necessity to leave her family and never return home again.

The family resemblance that exists between *Antigone* and the part of *Incendies* that so far has been narrated, also relates to the three words of honor, shame and necessity. The moral fight taking place between Antigone and Creon, her uncle and new king of Thebes, is a fight about which laws and honors come first; the laws made by gods or man-made laws. Creon advocates obedience to the man-made laws of power, while Antigone insists on the higher laws of duty to the gods, and that her duty to preserve the family honor outweighs her duties to the state. The value conflict between them as dramatized in the play can be summarized thus.

The battle of Thebes has just ended. The Argives were defeated and the corpses of Eteocles and Polyneices, Antigone's and Ismene's brothers, lie on the battle field. Creon, great uncle and uncle to both men, and brother of Jocasta, the natural mother and wedded wife of Oedipus, who has seized the throne due to his twofold blood connection with the dead, reads a proclamation announcing the fate of the bodies of the two brothers. Eteocles, who died defending the walls of Thebes, was to be given a full and honorable burial and his body was to lie in state in the Palace. Polyneices, who died spilling Theban blood by provoking mutual fratricide, was to be denied burial; his corpse was to remain out in the open at the mercy of birds of prey and savage dogs. And anybody who ventures to disobey the terms of this edict shall pay with his life (*Antigone*, lines 220–241). Antigone refuses to obey a man-made edict that goes against the higher laws of duty to the gods and to the dead, a duty she feels to be so strong that she in spite of knowing that her act of honor to Polyneices will lead to her death, she goes to sprinkle a fine layer of dust over his body, thus obeying Hades, the god of the underworld, who according to Antigone's belief "desires equal rites for both" (*Antigone,* line 593).

Back to *Incendies* and the role of death and honor. What Nawal states in the first part of her testament about the fate of her bodily remains is not a wish; it is a command issued to the notary that clearly echoes the rigid absoluteness and non negotiability of Creon's edict. As was the case with Polyneices she is not to be honored with normal burial rites. On the contrary, she is to be put in the ground, without prayers, naked and with the face turned towards the soil, without a coffin, and without a stone put on her grave or her name engraved anywhere. And her reasons for issuing this order: "No epitaph for those who do not keep their promises". This is, however, not the last word Nawal states about the fate of her bodily remains; as already mentioned, later in the testament she leaves to Jeanne and Simon the possibility of restoring her honor by tracing their father and brother and giving them the sealed envelopes: "When these envelopes have been given [...] The silence is broken. A promise kept. A stone may be placed on my tomb, and my name engraved on the stone in the sun".

In *Antigone*, Creon is warned by the old soothsayer, *Teiresias*, that disaster will afflict his house, and he himself will fall into disgrace in the face of his family and his people, if he insists on effectuating his edict:

> Then understand this well – you will not see the sun race through its cycle many times before you lose a child of your own loins, a corpse in payment for these corpses. You've thrown down to those below someone from up above – in your arrogance you've moved a living soul into a grave, leaving here a body owned by gods below – unburied, dispossessed, unsanctified. That's no concern of yours or gods above. In this you violate the ones below. And so destroying avengers wait for you, Furies of Hades and the gods, who'll see you caught up in this very wickedness. [...] It won't be long before in your own house the men and women all cry out in sorrow, and cities rise in hate against you – all those whose mangled soldiers have had burial rites from dogs, wild animals, or flying birds who carry the unholy stench back home, to every city hearth. (Sophocles 2014, *Antigone*, lines, 1192–1205, and lines 1207–1213).

Teiresias' dreadful prophecies frighten Creon back to his moral senses and he understands that he has to abide by the higher laws of duty to the gods by releasing Antigone from the rock tomb where she has been put to die, and by preparing a grave for Polyneices: "Alas – it's difficult. But I'll give up. I'll not do what I'd set my heart upon. It's not right to fight against necessity" (Antigone 1235–1237). But Creon's attempt at rectifying his wrongs has come too late. Disgrace and horror upon horror are in waith for him. Antigone who was to be wedded to his son Haemon has hanged herself in the cave where she was put. When Haemon – his only living son – who has become estranged from his father because of his edict (*Antigone*, line 871–875), finds Antigone hanged, he curses Creon and follows suit his bride by killing himself in front of his father (*Antigone*, lines 1358–1380). This leads finally, to the self-slaughter of Eurydice, Haemon's mother and the father's wife, who "with her last breath" curses Creon for being "the killer" of her "sons" (*Antigone*, lines 1450–1452). Honor, shame and necessity. Three morally charged words, which in the play demonstrate the potential of bringing a family to ruin and total destruction. How these three words determine the fate of Nawal and her family is acted out in the rest of the film and play and will be subject of attention in the next paragraph.

Shame, necessity and the tragic quest for identity.

The remaining parts of the film dramatize Nawal's transformation into a resistance fighter and her fate as a defender of war. In addition, Jeanne's and Simon's quest into the true identity of their mother, father and brother – and thereby themselves – occupies center stage. Recall the math professor's conversation with Jeanne: "You have come to learn that A, your father is alive; and B, you have another brother. What is ridiculous is that you are questioning what is inevitable. You need to know. Otherwise, you will never be at peace. And no peace of mind, no pure mathematics. It provides you a starting point".

The scenes shift to Daresh, and the arrival of Jeanne. She goes to the university where her mother has studied. The picture of the young Nawal provides her with the first piece of the jigsaw. She gets some help from a man of age. Jeanne shows him the old photo of Nawal with the Arabic letters written on the wall. The man comments to Jeanne: "She looks familiar. Perhaps she worked on the student newspaper. That was long time ago. The people who left changed their identity. But it's striking", he continues: "The photograph of your mother was taken in Kfar Ryat". He points to the Arabic letters in the photo written on the wall behind Nawal: It's the sign for Kfar Ryat. It's a prison in the South. You don't know of Kfar Ryat?". "No", answers Jeanne. "Then you're definitely not from here. You don't know the South".

A flash back scene follows from the offices of the university newspaper: Pictures of (student) demonstrations are shown, surrounded by armed forces. One photo displays Nawal in the middle. The sounds from several typewriters are heard and a woman dictating: "If the international community does not intervene immediately, the refugees crammed into camps along the border…..". Nawal is sitting at a table covered with papers, waiting for someone. Next, Nawal is addressing a woman: "We the students oppose the Nationalist party and the expulsion of refugees from the borders [in the South]". Another female student [Nawal's cousin] comments: "The Nationalists support the militias of the Christian right, who openly threaten the refugees". Nawal: "The refugees are armed now, and have the support of most Muslims in this country". The woman to Nawal: "But you yourself are Christian". Jeanne: "We stand for peace. This is not about religion". The conversation between the three women is interrupted when military vehicles are seen moving in the direction of the university campus. Nawal runs out into the street between the military vehicles. Her uncle runs out of the offices of the newspaper as well, pleading her to stay away.

"Nawal. Come back. Stay here!".

She shouts back: "Uncle Charbel, they shut the campus!".

THE UNCLE: There will be fighting. I want you to stay with me!

Nawal returns to the house of her uncle. The uncle has just received a telephone call informing him that Christian villages in the South have been attacked, and allegedly, that that was the reason why the university had been closed down. He is afraid that the fights will spread to Daresh and tells his family that they will go to the mountains till things have calmed down. Nawal's cousin addressing her father: "And the newspaper?" The father [Nawal's uncle]: "Don't worry. The paper can survive without us for a while". The daughter: "Ideas only survive if we're there to defend them!".

The next scene displays Nawal secretly leaving her uncle's house to travel south to trace the village and orphanage where her son was put. When she finally arrives, she finds the village empty of people and the buildings burnt down and completely destroyed. She is told by a man watching her from distance that all the Christians in the village have been killed by "Chamsedine and his men… to avenge the refugees", and that the Muslims in the village have fled to a camp in Deressa because of fear of the reprisals that will come. "Maybe", he tells her, "the children are with them. Maybe the children are in Deressa".

The merciless logic of reprisals

In the scene that follows Nawal is seen entering a bus full of passengers on their way to Deressa. Before entering the bus she hides the necklace with the cross and puts on a scarf to cover her hear. After a while Nawal falls asleep. When she wakes up she finds that the bus has stopped and she sees the driver arguing with a group of armed and masked men. Suddenly the men start shooting at the bus. Many of the passengers are killed or deadly wounded. Nawal is hiding on the floor together with a mother and her child. In the bus there is complete silence followed by a man climbing the top of the bus with a can full of gasoline which he starts pouring out on the roof to put fire on the bus. With the necklace and the cross in her hand lifted she shouts to the men: "I'am a Christian. I'am a Christian". One of the men drags her out of the bus. She turns her head towards the bus with the mother and the child and shouts: "My daughter! My daughter!". She grabs the child from the mother's arms and runs in the direction of the armed men who put fire to the bus by firing at it. When the bus catches fire the child runs back towards the bus and the mother but is shot before she reaches the bus. The men leave while Nawal is left alone with the bus on full fire.

In the next scene Jeanne is on her way to the south of the country to find the village of her mother. She wears her mother's necklace and listens to the war song that was played at the beginning of the film, *You and Whose Army?* She walks through the ruins of several houses before she arrives the house where her mother had been raised. She calls Simon in Canada to tell him where she is. She shows a man of age the old passport and photo of Nawal. He speaks only Arabic. The man orders a young boy to guide Jeanne to a house nearby where a group of women are gathered. One of the young women, Samia, speaks French. At first Jeanne receives a warm welcome and is offered tea. She starts talking with Samia, and mentions her family name, Marwal, whereupon Samia answers: "There are many Marwal here". Jeanne tells that she is Canadian and is looking for Suha. Samia seems a bit puzzled and points to her great grandmother. Jeanne continues: "I am looking for my father. His name is Wahab". She takes out her mother's passport: "My mother's name is Nawal

Marwan". She shows the passport to the old woman, who briefly looks at the photo and gives it away to another woman, who answers: "We do not know". Another woman is heard saying. "But we do know" followed by a "hush" from another woman. The women start speaking more and more loudly, and the woman who replied, "We do not know", shouts, "Her mother is a disgrace". The atmosphere in the room becomes more and more tense. Samia comments: "Marwan's family was struck with shame. The war came". Jeanne responds: "Uh, I am looking for Wahab". Samia translates the answer of her great grandmother: "She does not know Wahab". The great grandmother adds: "I cannot help you. Sorry, I'am tired". A young woman states something in Arabic and Samia translates: "If you're the girl of Nawal Marwan, you're not welcome here. Go home". The great grandmother starts speaking again. "Looking for your father, but you do not know who is your mother". These enigmatic words about identity recall queen Jocasta's warning words to Oedipus, the king in the play of the same name: "O you unhappy man! May you never find out who you really are! (Sophocles 1994, *Oedipus the* King, lines 1283–84). As was the case with Oedipus, in spite of the disturbing words warning of the old woman, Jeanne does not back off from the quest into the true identities of her father, unknown brother and thereby herself and her brother.

Didactic Task 4
According to Cairns it may "be more common in small-scale, face-to-face societies"to "construe one's experience in terms of shame rather than guilt" (Cairns 1993, *p. 46). "It is important, however", he continues, "to be clear what this statement entails: no absolute distinction between two types of society (for even the greater anonymity of modern society leaves considerable room for the development of a concern for one's image in the eyes of others); no distinction in terms of external versus internal sanctions; no exclusion of the phenomenon of conscience; and no denial of the existence of personally endorsed moral standards" (Cairns* 1993, *p. 46). In other words, experiences of shame and guilt have always been there; they belong to the world of humans. They represent cross-cultural universals.*

Although a sharp differentiation between a shame culture and a guilt culture probably is empirically untenable, the suggestion is to make use of this theoretical construct to analyse Jeanne's encounter with the women in the previous scene and reflect on the moral difference between a guilt construct and a shame construct of the same scene. In addition it is suggested to reflect on the scene with Nawal, Wahab, her brothers and the grandmother from the vantage point of shame versus guilt.

Memory, anger, revenge and payback

The next scene starts with a flashback sequence on Nawal and the bus on fire whereupon she is seen walking to Deressa on the search for her son. Shootings are heard continuously. She seeks refuge for the night in one of the shattered buildings in Deressa. Next, she is back in Daresh and finds the offices of the university newspaper completely destroyed. Then follows a sequence where she is talking with a resistance fighter.

NAWAL: I arrived after the massacre at the Deressa camp. Everything was smoldering. I searched for my son among pools of blood. I never want to forget what I saw and heard.
THE RESISTANCE FIGHTER: You say you oppose our enemy. That does not make you our friend. Why should Chamseddine trust you?
NAWAL: My son's father was a Deressa refugee. My son was swallowed by the war. I have nothing left to lose. I have only hatred for the Nationalists.
THE RESISTANCE FIGHTER: That's not what you wrote in Charbel's newspaper.
NAWAL: My uncle believed that his words and books would promote peace. I believed him.
Life taught me otherwise.
THE RESISTANCE FIGHTER: What do you want now?
NAWAL: To teach the enemy what life has taught me.

Pictures of the Nationalist leader on a wall with red crosses painted on the wall is displayed. Next Nawal is shown teaching French to the little son of the Nationalist leader in the home of the leader. One day she goes to teach the son she smuggles a pistol with her and fires two shots at the Nationalist leader. She is sent to the Kfar Ryat prison where her hair is brutally cut off before she is left alone in a tiny prison cell.

In the next scene Jeanne is on her way in a car to Kfar Ryat. A man showing her around in the former prison tells her: "Amnesty International repeatedly condemned the Kfar Ryat prison. This is the women's section. The cells are bigger. Some prisoners were kept for 15 years". Nawal asks: "When was it built". The guide: "After the massacres. Right after war started. They jailed 600 political prisoners". Walking through a corridor with cells the guide encourages Jeanne to enter one of the cells so that he can take a photo of her. Jeanne takes out the prison photo of her mother: "Do you know this woman". "No", answers the guide. "I don't know her. That was a long time ago, so I….". Jeanne interrupts: "Do you know anyone who worked here?"

In the following sequence, Jeanne is stopping her car to ask a man to guide her in the right direction. She parks the car in front of a building that seems to be a school. A man comes out and Jeanne tells him she is looking for a man called Fahim Harrsa. The man confirms his name and tells he is the janitor of the school. Jeanne asks: "And before?". "Before", he repeats chucklingly: "That was before I was school janitor. I've been the school janitor a long time". Jeanne shows him the prison picture of his mother, saying, "She was in Kfar Ryat". The man looks bewildered, and answers: "I'm the school janitor. End of story". Jeanne: "I need someone who knew her, who can tell me about her. She is my mother". Hesitantly he tells her: "She's the woman who sings". Number 72. She shot the leader of the right-wing Christian militias. They made her pay a high price. Very high, very high. 15 years". He continues briefly in Arabic before returning to French: "The woman who sings. She sang all the time. You're sure it's her?": I spent 13 years watching her. Have you ever spent 13 years watching someone? They did everything to break her. At the end she still stood tall and looked them in the eye. Never seen anyone like her. She wouldn't break. They were enraged. So they sent Abou Tarek". Jeanne: "Who's that?" A long pause follows before the man continues: "Abou Tarek. You know, sometimes, it's better not to know". Jeanne answers: "I live with it anyway. Go on". The man: "Abou Tarek…Abou Tarek, he was a torture specialist. He raped her again and again. To break her before her release. So she'd stop singing. In the end, she became pregnant. It was like that. I'll never forget. 72. Made pregnant by Abou Tarek. They waited until she gave birth in prison. Then they let her go". Jeanne at the verge of crying: "The baby? Did you see the baby". The man: "There was a doctor who sometimes came. I believe he went crazy. Others say he has a restaurant in Tel Aviv. That's bullshit. He went crazy or else he died. Now, the nurse…I know her. She lives in Daresh". Jeanne repeats "Daresh" and gives him a pen and her notebook for him to write down the address. The scene then switches to Jeanne back in Daresh having a heated telephone conversation with her brother in Canada: "Listen, Simon, I don't give a shit. Just shut up! Mom was in prison. Simon [Jeanne sighing, and crying] mom, mamma was raped. She had our brother in jail. I need you".

Didactic Task 5

In task 2 it was suggested to discuss the relation between anger and guilt. Another morally charged word associated with anger is revenge or payback. Nawal's conversation with the resistance fighter in the preceding scene alludes to the relation between anger and revenge, when she tells the resistance fighter that she wants to join the resistance so that she can "teach the enemy what life has taught" her.

In her book, *Anger and forgiveness*, Nussbaum questions the assertion among many philosophers and psychologists "that a wish for payback is a conceptual part of anger" (Nussbaum 2016, p. 22). She argues that the relation is causal and external rather than conceptual and internal. Nussbaum introduces a differentiation between three alternative ways of expressing anger; (1) as a reaction to offence against oneself, one's status and rank, conceived of as down-ranking; (2) as a reaction targeting the offender, implying revenge and payback; and third, anger as a reaction targeting the offensive act. But this expression of anger is forward-looking in a different way than payback anger since it aims at amelioration, not retaliation. She labels this form of anger "transition anger" and argues that it is morally superior to status anger and payback anger (Nussbaum 2016, pp. 17–40).

A suggestion is to discuss the relevance of this differentiation when viewing the film followed by the attempt at identifying sequences that dramatize these three different forms of anger. All three forms of anger are richly represented in the film.

Death is never the end of the story

"THE WOMAN WHO SINGS" displayed in red letters on the screen opens the next scene. Nawal is walking back and forth in the tiny cell. Guards bring a woman to a room with her head covered by a white sack. Then screams from the woman are heard. Upon hearing the screams Nawal starts singing.

Following Nawal's singing, a song in Arabic is heard and the scene shifts to Canada; first to the swimming pool on a wintery day, then to the flat of Simon. He is seen staring out of the window, with bags in his hands as if ready for a travel. Next, the scene shifts to the notary's archives and Mr. Lebel is heard talking to Simon:

MR. LEBEL: The contracts of notary Amyott. From 1868–1892. His son, Edouard, took over until 1925. All handwritten. My grandfather Charles. My father Henri. And me. The end, because I'm the last of the Lebels. I'll show you the will of a man who led parallel lives. Turned out he had three wives. Here, in Miami, and in Honduras, where he did business. So now, it's not three children involved, but eight. That was fun, believe you me! Look. Death is never the end of the story. It always leaves tracks. To find your brother, you have to retrace your mom's past".
SIMON: Jean, I just want to find my sister. That's all!
MR. LEBEL: I'm on it. I have a few things to wrap up. Pack your stuff, we leave in…My passport.
SIMON: We? Haven't you done enough?
Mr. LEBEL: Exactly, I can't pull out now. Grow up, Simon. I know you need me.
SIMON: We find her and come back.
MR. LEBEL: Promise. For a notary, Mr. Marwan, a promise is sacred.

A flashback scene follows with Mr. Lebel in the sick room of Nawal. She whispers something to him. Mr. Lebel writes something down in his notebook. The scene shifts to his office where Mr. Lebel is writing letters. On two envelopes is written "To the father" and "To the son" respectively.

The next scene opens with "SARWAN JANANN" written in red on the screen and takes place in the detention center. A guard puts a white cloth to cover the head of Nawal. She starts singing. Nawal is brought to the room of the torturer, who walks around her staring at her from all sides. She is handcuffed, but the cloth covering her head has been removed. As he leaves the room he orders her: "Now. Sing". The rape scenes are not displayed. Only Nawal lying on the floor after the rape has taken place is shown, followed by a sequence where a pregnant and crying Nawal is banging her belly as if to provoke an abortion. The guards handcuff her with her hands behind the back. Next, the scene shifts to the birth of two babies. A nurse is assisting her. Then a man and the nurse are seen carrying the babies towards a river to throw them away to die, as had been done with the babies of the other rape victims of the torturer. The nurse pleads the man to give her the babies: "Stop […] You did your duty. You threw them in the river. I'll look after them. They are the babies of the woman who sings".

Shameful identities – the echoes of Oedipus the King in Incendies

The babies who were carried to the river to be drowned but then rescued by the nurse recalls the fate of Oedipus the child in Sophocles play *Oedipus the king*. Shortly after his birth the child's father, king Laius of Thebes, ordered him to be killed, so as to escape the dreadful prophecy that he would perish by the hands of his own offspring (Sophocles 1994, *Oedipus the King*, lines 849–858). In the words of Jocasta, the king's wife and mother of the child, thus was the baby's fate:

> [...] before our child was three days old,
> Laius pinned his ankles tight together
> and ordered other men to throw him out
> on a mountain rock where no one ever goes
> (Sophocles 1994, *Oedipus the King*, lines 861-869).

As was the case with Nawal's twins the servant to whom Jocasta handed the baby to have him killed did not obey the order. Instead he gave him to a stranger to take him of to this own country (Sophocles 1994, *Oedipus the King*, lines 1413–1414). The stranger set the baby's ankles free and brought him as a gift to the childless King Polybus of Corinth. The king cared for him and raised him as if he were his own son. 'Oedipus', the name allegedly given to the child by fortune (Sophocles 1994, *Oedipus the king*, lines 1236–1237) means either "swollen feet" or "knowledge of one's feet", and recalls the scars that the piercing of his ankles had left on him. The three-dotted tattoo put on the right heel of Nawal's baby was meant to mark and safeguard his identity from oblivion, but in the same way as the scars on Oedipus's ankles came to represent a "dreadful mark of shame" (Sophocles 1994, *Oedipus the King*, line 1234), the tattoed heel of Nawal's baby bears witness of a frightening story about to be unrolled. So far, the film has made use of the tattoo twice, first, to secure the identifiability of the baby, second, to display Nawal's son on his way to become a child soldier. Later in the film the tattoo is displayed three times, and each time disclosing more of the horror and shame that was lying in wait for the house of Nawal.

In the next scene a car is seen driving through a tunnel in the direction of the center of the city Daresh with Mr. Lebel in the back of the car talking to another notary, Mr. Maddad. Simon sits in silence in front of the car. Mr. Lebel had asked assistance from Mr. Maddad to trace the father and brother of the twins. Simon is clearly annoyed, insisting that their task is to find his sister, nothing more, whereupon the notary Maddad answers: "Your sister's the easiest family member to find". Next, the scene switches to a hotel room with Simon and Jeanne embracing each other. She shows him the address of the nurse who assisted her mother during the delivery in the detention center: "We can go see her today". Simon answers: "Jeanne it's enough". Jeanne shouts back in an angry voice: "Just shut up and come with me. It's your job to find our brother. If not for her or yourself, do it for me". She forces him to go with her to visit the old nurse who is hospitalized. Next, the scene shifts to the sickroom of the old lady who is in sleep. The two notaries are present in the room together with Jeanne, Simon and a young nurse. The nurse whispers something in Arabic to the old lady. The local notary introduces Mr. Lebel, Jeanne and

Simon to her, mentioning the family name of the twins. Jeanne greets the old woman and thanks her for having received them. Jeanne asks her whether she knew Nawal Marwan. The young nurse repeats the question and the old lady is heard muttering something in Arabic. Jeanne continues: "You were a nurse at Kfar Ryat?". The young nurse repeats the question. Simon breaks his silence and asks: "Madame, we're looking for the child she had in jail. Can you help us?". The old lady confirms that she got to know Nawal Marwal while she was working in Kfar Ryat. She tells them she helped Nawal to give birth and that she rescued the BABIES and returned them to the Woman who Sings upon her release from the detention center. With a beaming smile and arms stretched out against them as if wanting to hug them she mutters their names in Arabic: "Sarwan! Janaan!". A face full of joy contrasts with the shocked faces of Jeanne and Simon, who suddenly has been made aware of themselves being the bitter fruits of rape and torture.

Next scene: Jeanne and Simon are seen in the swimming pool, immersed in their own thoughts and swimming ferociously back and forth in the pool as if trying to purify themselves from the stains of horror and shame left them at the core of their identity by their father. No words are uttered between them, but the embrace at the end of the scene speaks of mutual comfort.

Son, father, brother – when 1 plus 1 equals 1. The conjecture of Syracuse revisited.

The next scene is introduced with NIHAD written in red on the screen. Four young boys are seen running through a desolate street full of wrecked vehicles and remains of collapsed buildings while trying to hide from snipers. A gun is heard fired and through the binoculars of the sniper's rifle the bloody corpse of one of the boys is displayed. The back of the sniper, his young face and his feet are in focus. The three-dotted tattoo on his right heel reveals his identity: The lost son of Nawal and Wahab.

In the next scene the two notaries are talking together about the "Woman Who Sings".

MR. LEBEL: Nawal Marwan worked 18 years for me. As my secretary [...] But realize I didn't know her after all.

Next, the two notaries and Jeanne and Simon are having a meal in a restaurant. Mr. Maddad informs them that no death certificate of their father, Abou Tarek, has been found: "Many in this country disappeared. He may have gone abroad. That's what his kind often did. They're either dead or they left. Wait and see".

SIMON: Maybe this is the best.

MR. MADDAD: For the brother… Your client, Mrs. Nawal Marwan is from the village of Der Om. She gave the child to a midwife named Elham, who placed him at the Kfar Khout orphanage in May 1970. It was hard to locate the registries. Many documents were destroyed in the war. But I finally discovered the registries. The Sisters of Charity collected and donated them to the National Archives. In May 1970, the midwife of Der Om, Elham, gave up a male child for adoption. The only boy placed at the orphanage that month. Look, he was registered under the first name of Nihad. The dates match. It's him" [Both notaries are seen smiling]. MR. LEBEL [to Jeanne and Simon]: He's your brother. Nihad. [Jeanne and Simon are looking at each other].

MR. MADDAD: Kfar Khout orphanage. Nihad of May. That was his name for the records.

Nihad of the month of May. The month of arrival was a temporary surname. There were no adoptions at the time. The war had started. In 1974 the orphanage was destroyed. We must ask who destroyed Kfar Khout to learn what happened with the children. Between you and me, given what occurred in the region, there's little hope.

SIMON: He's probably dead, too.

MR. MADDAD: I didn't say dead. I said we lost his trace.

Simon [with a stern face]: They're both dead. We open the letters. Finished.

MR. LEBEL: I won't allow it. This kind of thing is sacred. It would be like a rape.

SIMON [with irony]: Great, it runs in the family. [Jeanne looks at him and sighs. Simon apologizes].

MR. MADDAD: There's one path left, but it's somewhat radical. The warlord who raised the orphanage is still alive. He may know what happened to Nihad and the others.

SIMON: It's been an eternity.

MR: MADDAD: You must understand, these years saw a series of reprisals that set off each other with merciless logic. Like an equation. A warlord is precisely the kind of man with a long memory. He'll remember, believe me. Wallat Chamseddine. That's his name.

SIMON: Where do we find him?

MR. MADDAD: You don't find him, he finds you. We can try something. Do you like tea, Mr. Marwan?

Simon looks quizzically at Mr. Maddad. Jeanne, with a smile, to Simon: "It's your turn".

In the scene that follows next a car is driving in a barren landscape. Mr. Maddad is heard saying: "You'll travel to the Deressa refugee camp. You'll drink tea. With whomever invites you. You'll tell them you are the son of the Woman who Sings. You're looking for Nihad of May. You can't go alone. I'll find somebody to accompany you".

The scene shifts to Simon and Mr. Lebel on their way in a car to the village of Deressa. The chauffeur accompanies them to a group of men and women in the village. They are invited to stay for tea by one of the men. After tea is served the man asks in Arabic why they have come to Deressa. The chauffeur translates. Simon answers that they are looking for a man called Nihad of May. The man answers he does not know anyone with that name, and he asks for the name of Simon. The man looks with suspicion at him. Simon tells that his mother comes from Der Om in the South: "My mother's the Woman who Sings". The man looks at him but he says nothing. A long silence between them follows.

In the next scene two men knock on the door of Simon's hotel room. Simon opens the door and the men tell they have come to offer help to Sarwan Marwan. Simon confirms his Arabic name. They say he will be back in one hour. Simon looks a bit nervous; he says he needs to change his shirt, closes the door and calls Mr. Maddad. He assures Simon it is safe for him to go with the men. The men accompany Simon to a car waiting outside the hotel.

A flashback scene follows next. Nawal leaves a car and enters a big, black car. A man speaks to her: "You helped us. Now it's our turn to help you. [Nawal's worn-out face is displayed]. You'll leave the country. It will be simpler. We'll find you a home and a job. I want you to have a better life. We have many contacts in America. You must go with your children. I know they're alive. Take them with you. Your children will help you, you'll see". Nawal keeps silent till the man mentions her children. With her eyes filled with tears she responds: "You can't ask that of me". The man: "Your children are our children. Your family is our family. Trust me. I will help you. I will always be there for you and your children".

The fate of the loved, but lost child

CHAMSEDDINE is written in red letters on the screen. Simon sits blindfolded in a room. Several men enter the room. A man greets Simon with his Arabic name: "Forgive the precautions, but it's for your protection. I am called Wallat Chamsedinne. Tell me, how can I help you, Sarwan Marwan?

SIMON: You knew my mother?
CHAMSEDDINE: Let's say she worked for me. How can I help you?
SIMON: I'm looking for a man named Nihad of May.
CHAMSEDDINE: Why do you want Nihad of May?
SIMON: He's my brother. My mother asked me to find him.
CHAMSEDDINE: How can you say that Nihad of May is the son of Nawal Marwan?
SIMON: My brother was placed at the Kfar Khout orphanage. We found the birth records.
The dates. Everything fits. Nihad of May is my brother.
[Chamseddine whispers something to one of the other men. All except Chamseddine and Simon leave the room. Simon is allowed to remove the blindfold].
CHAMSEDDINE: My friends and I attacked the Christians of Kfar Khout to avenge the aggressions against our brother refugees. There was an orphanage in Kfar Khout. I spared the children and I took them with us. Nihad was among them. We trained them, prepared them to fight alongside us. Nihad had a gift. He was special. He quickly became a formidable marksman. But he wanted to find his mother. He searched for months. I don't know what he saw or heard. He became crazed with war. He came back to see me. He wanted to be a martyr. His mother would see his photo on every wall in the country. But I refused. He went back to Daresh. He became the most dangerous sniper in the region. A real machine. He would shoot at anyone. And then there was the enemy invasion. And one morning, they captured Nihad. He'd killed seven soldiers. They didn't kill him. They trained him and sent him to the Kfar Ryat prison.

SIMON: He was in prison?
CHAMSEDDINE: Yes. As a torturer.
SIMON: Did he work with my father?
CHAMSEDDINE: No. He didn't work with Abou Tarek, your father.

The scene shifts to the hotel where the twins are staying. Jeanne knocks on the door of Simon's room.

JEANNE: You're okay?
SIMON: One plus one makes two.
JEANNE: What?
SIMON: One plus one makes two. It can't make one. [Jeanne looks worried at Simon and touches his skin].
JEANNE: You're feverish.
SIMON: Jeanne? One plus one…Can it make one? She looks at him intensely, and a gasp of surprise and horror slips through her lips when she suddenly understands the conjecture Simon is alluding to.

In a flashback scene Jeanne is seen with her mother on the way to the swimming pool. Jeanne points to an empty bathing chair: "Take it if you want". Nawal does not respond. They both take off their clothes and enter the pool. Nawal swims towards the edge of the pool. A group of men are seen standing close by talking in Arabic. Discretely from below she looks at the men with *the* expression of a 'déjà vu' in her face. The three-dotted tattoo on the right heel of one of the men is displayed. As if on the verge of fainting she manages to navigate herself towards the ladder of the pool and climbs the stairs. She walks slowly in the direction of the men and stops right behind the young man with the three-dotted tattoo. One of the other men addresses her: "Can I help you, madame?". The young man turns towards her, and they both stare at each other in silence, until Nawal responds, "I'm sorry ", and walks back to her chair. Her face expresses a combination of wonder and horror.

The film returns for a moment to the hotel room with Jeanne and Simon. Jeanne is in turmoil. She looks horrified and gasps and sobs while covering her mouth with her hand. Then two different landscapes are briefly displayed (barren hills versus a wintery landscape, probably symbolizing the Middle Eastern and Canadian roots of Jeanne and Simon) while the rest of the conversation between Simon's and Chamseddine is heard.

CHAMSEDDINE: When he became a torturer, your brother changed names. He became Abou Tarek. Nihad of May. We know he's living in Canada under a new identity. Nihad Harmanni. [On the screen the former torturer is displayed rolling a cleaning wagon in a hall full of busses or trams, followed by a sequence where he is cleaning the interior of one of the vehicles].

Lost and found. The silence broken. Honor restored

In the second last scene of the film Nihad is on his way home from work. Jeanne and Simon are seen walking towards him and Jeanne addresses him with his new name, "Mr. Harmanni". He confirms he is the man. They look at each other and Simon hands him an envelope, "This is for you". Mr. Lebel is waiting in a car, watching the encounter between Jeanne, Simon, and their brother/father. On entering the entrance of his apartment building Nihad is opening the envelope on which it is written, *Letter to father.* The letter reads:

> I'm shaking as I write. I recognized you. You didn't recognize me. It's magnificent, a miracle. I am your number 72. Our children will deliver this. You won't recognize them, for they are beautiful. But they know who you are.

He runs out of the building trying to find Jeanne and Simon, but they have already left. The reading of the letter, with the voice of the mother, continues:

> Through them, I want to tell you that you are still alive. Soon you'll turn silent…I know. For all are silent before the truth.
>
> Signed Whore 72.

Nihad enters his flat. Still standing he opens the second letter, written to the son:

> I speak to the son, not to the torturer. Whatever happens, I'll always love you. I promised you that when you were born, my son. Whatever happens, I'll always love you. I looked for you all my life. I found you. You couldn't recognize me. You've a tattoo on your right heel. I saw it. I recognized you. You are beautiful. I wrap you in tenderness, my love. Take solace, for nothing means more than being together. You were born of love. So your brother and sister were born of love, too. Nothing means more than being together. Your mother, Nawal Marwan, Prisoner nr. 72.

After the reading Nihad remains standing, as if paralyzed by the dreadful revelations in the two letters. He, who always had wanted to find his mother, who wanted to die as a martyr so as to make his mother know who he was, had been recognized by her, not as the heroic martyr, but as the vicious torturer and rapist, father to his sister and brother, but most important, the son born out of love.

The promise

In the last scene Mr. Lebel repeats the last phrase in Nawal's will and hands Jeanne and Simon the third letter: "When the envelopes have been delivered, you will be given a letter, the silence will be broken, a promise kept, and you can place a stone on my grave, and on it engrave my name in the sun". Simon breaks the seal of the envelope and takes out the letter.

Letter to the twins. It reads: "My loves, where does your story begin? At your birth? If so, it begins in horror. At the birth of your father? Then it begins in a great love story. But I say your story begins with a promise to break the chain of anger. Thanks to you, today I have finally kept it. The chain is broken. Finally, I can take the time to cradle you, to gently sing a lullaby to console you. Nothing means more than being together. I love you. Your mother. Nawal". The scene shifts to the graveyard and the stone on which is engraved, "Nawal Marwan. 1949-2009". In front of the grave the son born out of love is standing, paying his last respects to honor the mother he had always loved and been searching for.

Didactic Task 6

Based on the description below of 'moral injury' and the film's dramatization of moral injuries it is suggested to reflect on the following question: How can moral injuries in the wake of war be healed? A key message in the film seems to be that some sort of healing is possible if the chain of anger is broken and it is replaced by foregiveness and love. Then, maybe, peace in the mind and hearts of victims of war can be restored.

What is moral injury?

Like psychological trauma, moral injury is a construct that describes extreme and unprecedented life experience including the harmful aftermath of exposure to such events. Events are considered morally injurious if they "transgress deeply held moral beliefs and expectations". Thus, the key precondition for moral injury is an act of transgression, which shatters moral and ethical expectations that are rooted in religious or spiritual beliefs, or culture-based, organizational, and group-based rules about fairness, the value of life, and so forth.

Moral injury in war.

In the context of war, moral injuries may stem from direct participation in acts of combat, such as killing or harming others, or indirect acts, such as witnessing death or dying, failing to prevent immoral acts of others, or giving or receiving orders that are perceived as gross moral violations. The act may have been carried out by an individual or a group, through a decision made individually or as a response to orders given by leaders.

What is the aftermath of moral injury?

In terms of the aftermath of moral injuries, transgressive acts may result in highly aversive and haunting states of inner conflict and turmoil. Emotional responses may include:

- Shame, which stems from global self-attributions (for example "I am an evil terrible person; I am unforgivable")
- Guilt
- Anxiety about possible consequences
- Anger about betrayal-based moral injuries

Behavioral manifestations of moral injury may include:

- Anomie (for example alienation, purposelessness, and/or social instability caused by a breakdown in standards and values)
- Withdrawal and self-condemnation
- Self-harming (for example suicidal ideation or attempts)
- Self-handicapping behaviors (for example alcohol or drug use, self-sabotaging relationships, etc.)

In addition to grave suffering, these manifestations of moral injury may lead to under- or unemployment, and failed or harmed relationships with loved ones and friends (Source: Maguen and Litz 2016).

Didactic Task 7
Article 1 of the Universal Declaration on Human Rights states: All human beings are born free and equal in dignity and rights. They are endowed with reason and conscience and should act towards one another in a spirit of brotherhood.

Furthermore, Article 8 of the Universal Declaration on Bioethics and Human Rights emphasizes the importance of respect for human vulnerability and personal integrity.

In light of these two articles and the film's dramatization of the fate of women and children in the wake of war the suggestion is to discuss what moral integrity and dignity in situations of shaken and/or shattered identities might entail.

8.5 Some Didactic Afterthoughts and Concluding Remarks

In his famous definition of tragedy Aristotle highlights the importance of pity (*eleos*) and fear (*fobos*), in the sense that triggering these two emotions generates the kind of catharsis necessary for moral learning. The take home message for bioethics teaching of these poetic insights is that narrative representations – be they real or fictional – of this kind have the potential to promote the form of moral learning one should strive for, i.e. a kind of learning that appeals not simply to our rational capacities but impact our feelings and sensations of heart and body as well. As famously stated by Nussbaum in an early essay: "To be fully successful as arguments, arguments must be such as to change the heart" (Nussbaum 1990, p. 3). The film *Incendies* represents in this respect a paradigm case of what I believe Aristotle had in mind when he wrote the *Poetics*, i.e. dramatic representations of human upheavals and moral injuries that transcend the contingent constraints of historical accounts, thereby conveying a message of universal moral relevance. *Incendies* offers a narrative that in condensed form dramatize the human upheavals and moral injuries women, children and refugees are subjected to in the wake of war and armed conflicts. What can help to explain the killing of Nawal's boyfriend, Wahad, and the abandonment of her beloved child? And who is to blame? A culture in which shame and honour govern the minds and hearts of people? Nawal's brothers? The grandmother? As the old saying goes: "The strength of a chain depends on its weakest joint". What elements in this chain of events can help to explain the transformation of a child of love into a sniper and a torturer, or the decision of Nawal to become a resistance fighter? Who is to blame for the rape of Nawal? What can help to explain Nawal's silence in the face of Simon and Jeanne? The anger of Simon? Jeanne's feeling of guilt? Nahad's remorse? The impossibility of going home? Where does the story of the different victims of this war begin? What can explain and who is to blame for the merciless logic of reprisals so prevalent in the armed conflicts displayed? The film provides no definite answers to any of these questions. Instead it conveys the painful message that these are questions that need to be addressed again and again, because this is the nature of conflicts labelled ‘tragic’; when the work is done, moral twilight and ambiguity will still remain (Østerud 1976, pp. 75–76).

Finally, although no definite answer to these painful questions are offered in the film, an underlying message seems to be that when the rules that govern the ties of the social fabric have been destroyed – as often is the case in armed conflicts – the option that remains is the total instrumentalization of human life and the violation of fundamental human rights: the right to life; the right to liberty and security; the right to be protected from torture, and cruel, inhuman or degrading treatment or punishment; the right to freedom of thought and the right to education. If these rights are not in place there will be no possibility of maintaining a humane society, because then man by default becomes as a wolf to any other human being, and in particular to those most in need of protection; women and children. In such a society, governed by the merciless logic of wolves, there is no hope of finding permanent shelter, or a home where it is possible to rest, to grow and to flourish with dignity.

References

Aristotle. 1984. Poetics; nicomachean ethics; politics. In *The Complete Works of Aristotle. The Revised Oxford Translation*, ed. J. Barnes. Princeton: Princeton University Press.

Benedict, Ruth. 1946. *The Chrysanthemum and the Sword: Patterns of Japanese Culture*. Boston: Houghton Mifflin.

Cairns, D. 1993. *Aidos. The Psychology and Ethics of Honour and Shame in Ancient Greek Literature*. Oxford: Clarendon Press.

Developing Artists. 2013. *Queens of Syria*. Accessible at: http://www.developingartists.org.uk/our-projects/queens-of-syria:-jordan-&-uk-theatre-tour).

Dodds, E.R. 1951. *The Greeks and the Irrational*. Berkeley: University of California Press.

Doerries, B. 2015. *The Theater of War: What Ancient Tragedies Can Teach Us Today*. New York: Alfred A. Knopf.

Godwin, G. 2011. Introduction. In *You Can't go Home Again*. Wolfe T. New York: Simon and Schuster.

Hart, J., Tyrer, B. (2006). Research with Children Living in Situations of Armed Conflict: Concepts, Ethics & Methods. Accesible at: https://childhub.org/en/child-protection-online-library/hart-j-tyrer-b-2006-research-children-living-situations-armed

Holstun, J. 2015. Antigone Becomes Jocasta: Soha Bechara, *Résistante*, and *Incendies*. *Mediations* 29 (1): 3–42. Accessible at: www.mediationsjournal.org/articles/antigone-becomes-jocasta

Joe's Garage. 2010. *Movie Night: Incendies*. Accessible at: http://www.joesgarage.nl/events/movie-night-incendies

Konstan, D. 2012. Shame in Ancient Greece. Social Research 70 (4): 1031–1060. doi:https://doi.org/10.2307/40971960 (accessed December 2, 2012).

Maguen, S., and Litz, B. 2016. *Moral Injury in the Context of War*. National Center for PTSDU.S Department of Veterans Affair. Accessible at: https://www.ptsd.va.gov/professional/co-occurring/moral_injury_at_war.asp

Mouawad, M. 2010. *Schorched*. Toronto: Playwrights Canada Press.

National Alliance to end Homelessness. 2016. *The State of Homelessness in America 2016*. Homelessness Research Institute. Accessible at: http://endhomelessness.org/wpcontent/uploads/2016/10/2016-soh.pdf.

Nussbaum, M.C. 1990. Introduction. In *The Poetics of Therapy. Hellenistic Ethics in its Rhetorical and Literary Context*, ed. M.C. Nussbaum, 1–6. Alberta: Academic Printing & Publishing Edmonton.

———. 2016. *Anger and Forgiveness. Resentment, Generosity, Justice*. New York: Oxford University Press.

O'Mathuna, D.P., V. Dranseika, and B. Gordijn, eds. 2016. *Disasters: Basic Concepts and Normative Theories*. Dordrecht: Springer/Nature.

Ohly F. 1992. *The Damned and the Elect: Guilt in Western Culture*. Trans. Linda Archibald. Cambridge: Cambridge University Press.

Østerud, S. 1976. Hamartia in Aristotle and Greek Tragedy. *Symbolae Osloenses* LI: 65–80.

Radiohead. 2001a. *You and Whose Army?* Accessible at: https://www.google.com.uy/?gfe_rd=cr&ei=VBFVWbSIKaiB8QeZl4aADg&gws_rd=ssl#q=youtube,+You+and+Whose+Army?&spf=1498747223738.

———.2001b. *You and Whose Army?* Lyrics. Accessible at: https://www.google.com.uy/?gfe_rd=cr&ei=y-hHWd3XBoLX8gfLq47QCQ&gws_rd=ssl#q=you+and+whose+army+lyrics&spf=1497884875799.

Solbakk, J.H. 2006. *Catharsis* and Moral Therapy II: An Aristotelian account. *Journal of Medicine Healthcare and Philosophy* 9: 2141–2153.

———. 2015. Movements and movies in bioethics: The use of theatre and cinema in teaching bioethics. In *Bioethics Education in a Global Perspective*, ed. H. ten Have, 203–221. Dordrecht: Springer.

Sophocles. 1994. Ajax; Electra; Oedipus Tyrannus. In *Sophocles I*, ed. H. Lloyd-Jones. Cambridge, MA/London: The Loeb Classical Library, Harvard University Press.

———. 2014. *Antigone.* Trans. Ian Johnston. Accessible at: https://records.viu.ca/~johnstoi/sophocles/antigone.htm.

Their World. 2017. *10 Countries Where Child Soldiers Are Still Recruited in Armed Conflict*. Accessible at: http://theirworld.org/news/10-countries-where-child-soldiers-are-still-recruited-in-armed-conflicts).

UNHCR. 2015. *Global Trends. Forced Displacement in 2015*. The UN Refugee Agency Accessible at: http://www.unhcr.org/statistics/unhcrstats/576408cd7/unhcr-global-trends-2015.html.

Williams, B. 1993. *Shame and Necessity*. Berkeley/Los Angeles/Oxford: University of California Press.

Chapter 9
To See Differently: Incorporating the Arts into Bioethics Education

Terry Maksymowych

Abstract Traditional approaches to teaching bioethics courses include, among other things, case studies from clinical or research ethics, chapters on patient autonomy and beneficence, conflicts of interest and the dangers of hidden prejudice. All of these are essential to a good bioethics course. In my experience, however, incorporating the arts—whether it is a poem, short story, film, or a painting or play—enhances the curriculum and encourages students to see differently. The arts have the power to startle, to challenge beliefs, to invite people to see the world through others' eyes. In our culture, the arts are often relegated to the province of entertainment, but they can be powerful teaching tools, particularly in a field of study in which empathy is highly valued. Every culture has stories, art, music and dance, and these can be used to speak to students in introductory or advanced bioethics courses. In science and humanities courses, they can be used to introduce bioethical dilemmas and promote discussion. An instructor can use examples from the students' own culture and expand their experience by introducing the arts and particular ethical issues of other cultures. This chapter will focus on the methodology of incorporating the arts into bioethics classrooms, using examples from various cultural traditions.

9.1 Introduction

Traditional approaches to teaching bioethics courses include, among other things, case studies from clinical or research ethics, chapters on patient autonomy and beneficence, conflicts of interest and the dangers of hidden prejudice. All of these are essential to a good bioethics course. In my experience, however, incorporating the arts—whether it is a poem, short story, film, or a painting or play—enhances the curriculum and encourages students to see differently. The arts have the power to startle, to challenge beliefs, to invite people to see the world through others' eyes.

T. Maksymowych (✉)
Academy of Notre Dame de Namur, Villanova, PA, USA
e-mail: tmaksymowych@ndapa.org

H. ten Have (ed.), *Global Education in Bioethics*, Advancing Global Bioethics 10,
https://doi.org/10.1007/978-3-319-78984-2_9

In our culture, the arts are often relegated to the province of entertainment, but they can be powerful teaching tools, particularly in a field of study in which empathy is highly valued.

9.2 Poetry and Patient Autonomy

The beautiful and haunting poem, "Monet Refuses the Operation" helps students see the world from a disabled patient's perspective, where quality of life may be interpreted quite differently from that of the physician's usual experience (Mueller 1996). Monet developed cataracts, which became progressively worse through his later years. His art mirrored his view of the world as he grew older and many people appreciate the effect of blurred borders and running colors. In the poem, author Lisel Mueller imagines a conversation between the artist and his doctor, where Monet expresses the insights he has gained from his living with his disease. In my experience, everyone from high school bioethics students to physicians coming for a continuing education workshop are shocked and silenced as I read the poem aloud. There are usually several minutes of silence after I finish and that silence is filled with the wonder of seeing differently. Instead of disease and disorder being solely something to be eliminated, with frustration when a patient does not agree, students and healthcare professionals realize that perhaps the illness or injury means something very different to the patient. Instead of begrudgingly allowing for patient autonomy, a newfound respect for patient choice is achieved.

9.3 Short Story and Physical Enhancement

Hawthorne's short story "The Birthmark" illustrates the dangers of scientific research without sensitivity to the needs and wellbeing of the patient, as well as the themes of physical enhancement and perfectionism (Hawthorne 1987). Students often assume that distrust of science is a new phenomenon and that obsession with physical enhancement is unique to their generation's reality television shows. This short story describes a would be scientist in the early nineteenth century who is determined to obliterate a small birthmark on the face of his otherwise perfect wife (it is interesting to point out to students that he doesn't become obsessed with her perfection until after she becomes his wife). Although at first confused by the intensity of his drive to find a scientific way to cure her, the wife becomes obsessed over time with freeing herself of the birthmark, even after finding evidence that not one of her husband's experiments have ever been successful. Eventually, the experiment succeeds at ridding the wife of the birthmark, although she dies moments later. The themes of fear of irresponsible science, the drive for perfection through physical enhancement, and the social power to determine what beauty means are all ones that students can identify with and can find many examples in society today of the same struggles. For those students who do not find a sense of identification with physical enhancement, introducing the topic of mental enhancement usually brings them to the discussion with more enthusiasm. Students who have no patience with endangering their health for sports often feel quite differently when the issue is SAT scores and grades.

9.4 Television and Human Experimentation

The HBO movie "Miss Evers' Boys" puts faces and personalities to those physicians, nurses, and patients involved in the infamous Tuskegee Study (Cobb 1973). Case studies have been used extensively in teaching bioethics because they provide names and faces to the diagnosis or dilemma. A film, play or television show takes this experience several steps further. In the case of "Miss Evers' Boys," the movie tries to be true to the historical facts, although some characters were enhanced and some license is taken for dramatic effect. Discussion with students before or after about these differences can stimulate an exchange of viewpoints on the relationship between historians, scientists and the film industry. "Miss Evers' Boys" also has a scene that is particularly helpful in illustrating how people with good intentions can be compromised ethically over time. The main character, Miss Evers, is arguing with her superior over the decision to no longer treat the men in the study in the hopes that federal money will be forthcoming for treatment in the future. She describes the situation of "being taken up over the hill," where one is encouraged to make a compromise to overcome an obstacle, only to find another compromise is needed for another obstacle, and on and on. After a time, the person finds themselves severely compromised in a situation they would never have agreed to at the beginning of the journey. This experience can also be illustrated by the popular television series "Breaking Bad," where a popular high school chemistry teacher is given a fatal diagnosis and begins to sell drugs in order to provide for his family's future. He is a very sympathetic character, but by the end of the series is nearly unrecognizable in his descent into illegal and immoral acts. No one who hadn't watched the entire series would be able to find any sympathy for the character. These examples are helpful in trying to address students' questions about how normal people can do horrible things in ethics units featuring human experimentation topics.

9.5 Film and Genetic Engineering

The classic Hollywood film, "GATTACA," explores the issues of genetic engineering and privacy. Although many scientists do not appreciate the dystopian view of a future where biological determinism is the norm, fearing that it will fan the fire of distrust in science, the film is actually a study in how genetics do not exclusively determine the future, that they can only describe predispositions. It is an excellent opportunity to discuss the role of environment on gene expression and the limits of the genetic screening of embryos. The movie's main character is identified as a "de*gene*rate," a person whose parents did not take advantage of preimplantation genetic screening and engineering. In the futuristic society portrayed in the film, such individuals are relegated to menial labor, producing a caste society where genetic predisposition is converted to determinism. The film focuses also on the lack of genetic privacy, where a drop of blood, saliva or a strand of hair can be easily tested for everything from career to dating prospects. The main character refuses to accept his second class citizenship and succeeds, after many dramatic struggles, in attaining his dream of being an elite astronaut. Students are interested in the fact that much of the genetic technology in the movie is available today and the rest probably will be accessible in the near future. Discussions of how to utilize the best aspects of genetic engineering while condemning that which is divisive and immoral are ensured.

9.6 Various Media and Disability Ethics

The paintings of Van Gogh and O'Keefe, the poetry of Coleridge and Byron, the novels of Woolf and the music of Schumann, offer glimpses into the minds of creative geniuses struggling with mental illness. The Academy Award winning movie "A Beautiful Mind" follows the descent into schizophrenia of Nobel mathematician John Nash. This is also an opportunity to discuss the reality of Nash's experience of mental illness with the "Hollywood-ized" version. The documentary "People Say I'm Crazy," which follows the artist John Cadigan through a much more realistic and harsh experience with schizophrenia can be a helpful contrast to the Nash film. The study of mental illness can lead to bioethics discussions of mental diversity, forced treatment, pharmacological versus behavior modification therapies, and how cultures through history and around the world view mental illness. Including a discussion on disability ethics is also helpful here, as it is in any discussion of biotechnology. Students are often unaware of the malleability of decision-making about what constitutes a disability in a particular society. The subject of disability ethics startles students. The concept that a particular time and culture "decides" what is a disability encourages them to think "outside the box." The idea that disability is a cultural construct is challenging and enlightening to students. To ask them, "Who decides?" opens conversations about not only disability, but genetic engineering and other mechanisms that aim to rid the world of disease and disorder. A study of the American eugenics movement shocks and saddens students. Most textbooks and history curricula do not attend to this cultural phenomenon of the early twentieth century. It had a profound effect, however, on the Nazi program as well as much human experimentation in the United States. Also, it can inform students about the current medical efforts to erradicate genetic disorders that science and our society deem dangerous. Are the present genetic and reproductive technologies a new version of eugenics? These questions stimulate student research and debate and history becomes real and significant to twenty-first century youth.

9.7 Photography and Gender Ethics

The arts are rich in illustrations of bioethical issues in the fields of gender and LGBT studies. Del LaGrace Volcano (see www.delagracevolcano.com) is a formally trained photographer who uses that medium to confront binary gender culture and highlight the damage done to the bodies of intersex individuals through unnecessary surgeries. For more than half a century, infants born with intersex conditions were assigned a gender and surgically altered to match that gender. Parents were not given information because the medical community felt that it would be a burden to them and may hinder their ability to raise their child unambiguously. The prevailing thought at the time was that gender was the result of nurture rather than nature and that children raised in an assigned gender would conform to that identity. However, late in the twentieth century, adult patients with gender dysphoria and side effects of surgery including scarring and lack of sensitivity, began to advocate for interventions to be postponed until the child could participate in the decision-making. Although the new medical protocol is now to wait, there are still many physicians and surgeons who do not follow these recommendations. While they now inform parents about the condition, they continue to use language, facial expressions and misleading promises to encourage parents to consent to the surgeries. While most students are aware of many LGBT issues, most do not know the history of intersex treatment. Some of the most important clinical ethics values can be illustrated by studying this history. Volcano's photographs as well as published stories written by patients create empathy and a better understanding of the need for medical ethics.

9.8 Theater and Sexuality

The play "The Laramie Project" and the film "Philadelphia" can be used in the classroom to help students understand the fear and prejudice homosexuals have faced in America in the latter part of the last century. "The Laramie Project" chronicles the events in Laramie, Wyoming in October of 1998, when two young homophobic men lured a gay student, Matthew Shepard, away from a bar and beat him to death. The play is a collection of quotes from many of the principals involved in the crime and court cases that followed, as well as interviews with townspeople, clergy and medical personnel facing their own reactions to the crime. Since violence against members of the LGBT community is still common, this play—later made into a TV film—stimulates discussion of the ethical issues involved. The Academy Award winning movie "Philadelphia" is set a decade or more earlier than the Shepard case and is the story of a successful Philadelphia attorney who is unjustly fired from his law firm when the senior partners learn that he is gay and HIV positive. The film is based on a true story and helps to show students the fear that permeated our society when HIV/AIDS was a largely unknown and misunderstood disease. In a biology class studying HIV, this film, whether shown in class or assigned for home viewing, can put human faces on this devastating disease. The disease is no longer a diagram in a textbook, but a force that destroys human lives and raises important questions about our response to it. This can also be a springboard for ethics discussions about patient care, surrogate decision-makers, pharmaceutical research and costs, government grants for study, and insurance.

9.9 Creative Writing in Bioethics Class

Engaging students in their own art expressions can be an extraordinary teaching experience. In my classes, students are challenged to expand their thinking and encouraged them to "take chances" with their own creativity. For example, after studying the Nazi doctors' human experiments (not for the faint hearted), I assign my students to imagine themselves in a South American café where they encounter the Nazi "Angel of Death," Joseph Mengele, who escaped to South America after the war. What would you ask him? What would he reply? These questions help students to move out of their cultural "bubbles" and engage with persons who may think *very* differently from themselves. I ask the students to write a dialogue between a student of bioethics and a Nazi scientist. This assignment initially draws groans and great trepidation, but eventually leads to a greater understanding of the complexities of engaging with people who are ideologically opposed to you. Another assignment that initially is scary but leads to amazing results is one associated with the study of schizophrenia and the film "A Beautiful Mind." In the film, John Nash has delusions and hallucinations that feed his personal fears and aspirations. For my class, after analyzing this aspect of schizophrenia, The students are asked to think deeply and imagine that, if *they* were schizophrenic, what delusions and hallucinations they would have and why. This assignment has produced a great deal of soul searching and discussion with family and friends. My argument for such an exercise is that bioethics involves self-knowledge as much as empathy for others. This kind academic work strengthens the realization that individuals do not always know what "matters" to them, and the role of the clinical or bioethicist is to facilitate such knowledge.

9.10 Incorporating Bioethics into the Classroom

Every culture has stories, art, music and dance, and these can be used to speak to students in introductory or advanced bioethics courses. In science and humanities courses, they can be used to introduce bioethical dilemmas and promote discussion. Bioethics is timely and engaging for students. It encourages critical thinking, problem solving skills, encourages scientific argumentation and literacy, and enhances the appreciation for open discussion and empathy. Case studies can be used to capture the attention of a class when beginning a unit or in the midst of a unit that is not particularly interesting to students. A study of the skeletal or muscular system becomes more exciting when the topic of performance enhancement in sports is introduced. If the bioethics of a society's response to an epidemic is incorporated into a chapter on viruses or bacteria, students find the scientific material important and relatable. Every biology instructor has been assailed with bioethics questions when teaching genetics and biotechnology units. However, the biology classroom is not the only one where bioethics can be a welcome addition. Chemistry classes with the ethics of pharmaceutical regulations or astronomy classes with the inclusion of the ethics involved in genetically altering astronauts for life on Mars are other examples from science. The bioethics of human experimentation can be included in history, psychology and sociology courses. Literature and journalism courses can be enriched with works that have a bioethics theme. Theology instructors have a multiplicity of options for the inclusion of bioethics topics.

9.11 Concerns Over Time and Controversy

From middle school instructors to medical school professors, I have heard renewed excitement and enthusiasm for education when teachers have incorporated bioethics into their classrooms. The two initial concerns I hear at workshops and presentations to educators are time constraints and discomfort with controversial issues. These considerations, while valid, can be managed with careful planning and networking with teachers like me who have been incorporating bioethics into courses for decades. I argue that the enthusiasm and engagement level exhibited by students doing bioethics enhances learning and the skills developed can easily be transferred to other course material. A bioethics component can be done in one class or as a whole course, or anything in between. When teachers see how successful this is, they are encouraged to continue to use precious instruction time in this way.

Although some bioethics topics, such as genetic engineering and end-of-life decisions, can be very controversial, some teachers can convince administrations, boards and parents that it is better to arm students with the skills to evaluate arguments, engage in respectful discussions, and discern their own ethical values before they are sent off by themselves to colleges and universities where they will find others with very different values and opinions. If this approach does not work, there are many bioethics topics that are less controversial and would be appropriate for earlier grades or more conservative schools. Performance enhancement is one such issue, since it would be odd for those in authority to argue against a student learning about this. Another example of a less controversial topic would be environmental bioethics, providing there is no particular issue in a community that might be struggling with this. Even in my class of seniors, we generally move from "low-impact" topics in order to learn analytical and argumentative skills before progressing later in the year to more "high-impact" issues.

9.12 Introducing the Global Component

Introducing a global component to bioethics education is less intimidating than most teachers imagine. Learning that other cultures have different views on the study and implementation of bioethics is crucial for student success. Whether it is in traveling to another country or working with people of other cultural backgrounds in our towns and cities, students need to understand and appreciate diversity. An emergency room physician or nurse, for example, must be aware of differences in the understanding of patient autonomy or informed consent when faced with patients who do not share her traditional Western interpretations. While there is much common ground, a show of intolerance or frustration when presented with difference can cause damage to the medical professional-patient relationship. The same is true when pursuing business dealings or any other interaction that requires cultural sensitivity.

While most clinical ethics classes—and business ethics classes—address cultural differences and the need for understanding and empathy, it is usually done through a case study or two before the class moves on. In clinical ethics, generally there is a case study involving Jehovah's Witnesses and blood transfusions, one on fully informed consent with traditional Navajos, and perhaps one addressing how to respond if a patient wants a traditional healer consulted. If the course is being taught in a Roman Catholic university, the particulars of Catholic bioethics are usually taught. Generally, students respond with a frown at the "peculiar" beliefs of some people and move on. There has long been a discussion between clinical ethicists: How do we teach empathy? Exposure to other cultures, whether in medical or nursing school or on the job, seems a logical prescription. However, in my view that approach is too little, too late. In my experience, exposure in this way makes students who are already empathetic more so, but does little for the less empathetic student.

The growth of medical humanities courses in universities promote the concept that the arts enhance learning, observational skills, and empathy. It is my assertion that incorporating the arts into bioethics education, starting in the early grades and continuing into college and professional schools, is key to enhancing empathy and cultural sensitivity. This can be done with the technology available in most schools and utilizing the rich diversity of cultures already present in many classrooms.

9.13 Methodologies for Global Bioethics Study

In a classroom of students with a diversity of cultures and religious traditions, a bioethics educator can take advantage of this wealth of experience to address how different cultures or religious backgrounds might view patient autonomy, for example. In my classroom, students from Africa, China, India and Saudi Arabia have enriched our discussions tremendously. For example, during a unit on animal rights, my student from Kenya was able to contrast the differences she saw between the Western sympathies for African animals and antipathy toward trophy hunting and poaching with her own culture's view of animals as resources at best and dangerous predators at worst. While it is important to emphasize that one or a few students do not represent the attitudes of all people in their country, this method can help generate empathy and sensitivity toward others who do not share the same point of view. A student in a country that has age-old traditions of male hierarchy and respect for elders may be quite confused by the Western emphasis on individual autonomy. In most parts of Africa, for example, songs and folktales that have been passed down through generations warn youngsters of the dangers of disobeying parents or tribal elders (see, for example, anikefoundation.org). Another common African value that finds its way into their arts is the interwoven nature of the environment and humankind, as well as the interwoven nature of one generation with all those that have come before and those that will come after. These visual and performing arts can help Western students better understand the unique way that Africans develop their own approach to bioethics. In Africa, studying these same arts enable young people to articulate their values, examining what areas of Western bioethics is useful to them and which are not. Another example would be the poetry of A.K. Ramanujan (see www.poetryfoundation.org/poems-and-poets/poets/detail/a-k-ramanujan), which often highlights the continuity between the plant kingdom and the human body, where all the saps of the plant world infuse the human body, ensuring its unctuousness and fluidity. The belief that that all living things have personalities and rights can be found in many Indian art forms. The impact on bioethics discussions (for example, the use of alternative medicines) in India and around the world can only be imagined.

A particular piece of literary, visual, or performing arts from one tradition could be experienced and analyzed, with students from other traditions encouraged to explain how their culture might differ. Encouraging students and their parents or guardians to offer examples of arts in their cultures that would highlight these differences would be an effective method of exposing students to different backgrounds without the teacher having to be an expert on all traditions. Adults modeling acceptance and respect, as well as interest and appreciation, would, if done consistently, imbue children with these values.

In addition to this technique, or in the case of a more homogeneous student population, technology can be utilized to bring different cultural traditions together. With the widespread use of the internet and social media, educators are able to connect with people all over the world, most of whom are happy to share their cultural traditions with others. A teacher may use this connection for examples of popular artistic expression to use in the classroom. A live or recorded interview with people a world away—both adults and young people—can enable students to experience other cultures without the expense of visiting many other places. Exchange or study abroad programs are wonderful opportunities for students, but can be very expensive and are limited in ways that daily internet access is not. Allowing students to share examples of popular television shows or poems, music or paintings they feel illustrate a truth about bioethics norms in their own culture is engaging, challenging and promotes cultural appreciation and empathy. Students gain a deeper understanding of how the arts influence a culture's views on topics that matter to them. Students will learn that some of the most hotly debated ethical dilemmas in their own country may not be an issue at all in another. For example, the concerns about noninvasive prenatal technology may not be a primary concern to persons where basic access to care is in question.

An instructor can use examples from the students' own society and expand their experience by introducing the arts and particular ethical issues of other cultures. Through mining the wealth of diversity in the classroom and in a global classroom, the teacher need not become an expert on the arts in their own or other cultures before engaging students in an exploration of bioethics and the arts. The arts capture our human imagination and allow us to see from another's point of view—to see differently. The arts by definition expose us to the artist's own vision of reality and the appreciation—not just tolerance—of the multiplicity of viewpoints is key to a global bioethics.

9.14 Conclusion

In conclusion, the visual and performing arts can add depth and richness to bioethics classrooms. They can be used to better understand the values of other cultures as well as helping students articulate and examine their own values. While ethics classes can be found in most courses of study in colleges and universities, this is simply too little, too late. Whether a student is interested in business or medicine, sensitivity to other people and their cultural values is paramount to successful outcomes. Waiting until young adulthood to introduce ethics discussions is largely ineffective. Ideally, ethics—and particularly bioethics—should be introduced with age-appropriate topics early in education. Students find it interesting, timely and immediately useful to them. Empathy can be encouraged through exposure to other cultures and divergent points of view within cultures. The arts startle us out of our own point of view, challenging us to see differently, through others' eyes. It is in a culture's art that one can clearly see what that culture values, what is important to the people who make and appreciate that art. Experiencing the art of another culture helps students to see through others' eyes much more effectively than through reading facts about a culture in a textbook. It is important for a medical professional, for instance, to understand the rules and regulations associated with clinical ethics. But to truly understand the patient's experience, narrative or poetry or film or photography can be powerfully worthwhile. In addition to helping students in one culture better understand those of others, the arts can help students to identify and analyze their own values. Why does this poem make one cry, or that painting make one angry? Understanding one's values is imperative in crafting bioethics, but that understanding can require a great deal of unpacking. The arts' ability to shock and startle, to force us to see differently, whether one is in grade school or gradute school, argues for their incorporation in all classrooms.

Films

A Beautiful Mind, dir Ron Howard, 2001, DVD.
Breaking Bad, cr Vince Gilligan, 2008-2013, TC series.
Gattaca, dir Andrew Niccol, 1997, DVD.
Miss Evers' Boys, dir. Joseph Sargent, 1997, DVD.
People Say I'm Crazy, dir John Cadigan, Katie Carigan, 2003, DVD.
Philadelphia, dir Jonathan Demme, 1993, DVD.
The Laramie Project, dir Moises Kaufman, 2002, DVD.

References

Cobb, W. 1973. The Tuskegee syphilis study. *Journal of the American Medical Association* 65 (4): 345–348.
Hawthorne, N. 1987. *The Birthmark*. New York: Penguin.
Mueller, L. 1996. *Monet Refuses the Operation. Second Language*. Los Angeles: Louisan State Press.

Chapter 10
The Cinema as Instrument for Teaching Global Ethics

Flávio R. L. Paranhos

Abstract The use of movies as a tool for teaching ethics and bioethics has the virtue of drawing attention to several different issues, through an instrument of entertainment. With the help of the internet, one has access to a variety of interesting films, some of which are presented here. As it is common with movies, many of them are not original productions, and are based on another literary medium such as a play or a novel. That is the case of *Death and the Maiden, An Enemy of the People* and *The Trial of God*, which are originally plays, and *A Clockwork Orange*, which is a novel. Of the cases presented here, only *Decalogue* was originally intended as a movie. This only makes them even more intellectually rich, giving us the opportunity to discuss themes with global appeal such as Kantian versus utilitarian moral choices, free will, the divine beneficence, or even the role of a professor of ethics. The result is a very efficient way to engage the students' attention, the first step towards a successful course of ethics.

10.1 Introduction

The teaching of ethics has certain peculiarities that, while presenting themselves as difficulties, can also be an exciting challenge. The main concern of such ethics course is what exactly is expected as a result of it. If ethics is what ought to be done, those who study ethics will become experts on that which ought to be done. Certainly, things do not work quite that way. What a teacher of ethics can aspire is to nurture the critical spirit of the students in such a way that they will be able to never conform to what is given. Such disposition of spirit is even more important if we consider ethics globally. Students and teachers must do the exercise of questioning things as if they are not just in their backyard (or in a hospital ward), but in the entire planet. The position of those who are dedicated to the teaching and learning of ethics should never be comfortable. Following the existentialists, one needs to

F. R. L. Paranhos (✉)
Pontifical Catholic University of Goiás, Goiânia, Goiás, Brazil
e-mail: flavioparanhos@uol.com.br

H. ten Have (ed.), *Global Education in Bioethics*, Advancing Global Bioethics 10,
https://doi.org/10.1007/978-3-319-78984-2_10

take positions, assume responsibility for what happens in the world. In this sense, the arts and humanities are powerful instruments to sharpen the critical spirit of both students and teachers.

The use of movies as a tool for teaching ethics and bioethics has the virtue of drawing attention to several different issues, through an instrument of entertainment. With the availability of cheap DVD movies and even those freely and legally available on the internet, students can watch them at home before the scheduled class for a given subject, with the support of suggested literature. Then, during class time, rich discussions will occur, which will help the teacher to successfully engage the students' interest in key concepts in ethics and bioethics.

10.2 What Does a Professor of Ethics Teach?

A good movie to start a course on ethics would be *Thou Shalt Not Bear False Witness*, which is part of *Decalogue*, by the Polish director Krzysztof Kieslowski, composed of ten episodes inspired by the biblical ten commandments.

A Jewish researcher living in the United States was a translator of a known Polish ethics professor. One day, the researcher visits the professor in Poland and asks her to attend one of her classes. During class, Kieslowski presents us with a quote of himself, made in a rather ingenious way. A student presents a case to the professor, a moral dilemma. This is precisely the case of the second *Decalogue*, where a doctor is molested by his neighbor who wants to know if her husband has any chance to live (if he is likely to live, she will abort the child of her lover, if not, she will keep the pregnancy). The professor smiles condescendingly, because she already knew the case, because the protagonists were her neighbors.

The researcher, who was just observing, decides to intervene. She presents another example. During World War II, a Jewish girl had to be taken out of the ghetto and separated from her parents. She needed to be baptized and adopted, otherwise she was at risk of following the fate of her parents. A good soul arranges for a Catholic family to carry out the plan. However, when they arrive at the family's home, the mother just offers them tea, and the father does nothing. They had changed their minds, supposedly because it would be a sin "to provide false testimony".

If that would have been really the case, we'd have a strong evidence of the failure of a dogma camouflaged as categorical imperative (do not lie). Such a rationale would not only be defective, it would be unacceptable.

It does not take long for Kieslowski to deconstruct it, by the mouth of one of the students, who suggests that it was in fact the fear, not a supposed coherence (paradoxical coherence, it is true), that made the couple of devout Catholics decide not to help the little girl. Well, this little girl is now in front of the woman who denied her refuge, ironically in the middle of a class of ... ethics.

Evidently, the professor recognizes that the story is actually a story where she is the protagonist. She then dismisses class and meets the researcher in the Hall of the University, where she confirms what she suspected, and invites her to dinner. On the way home, the professor stops by the House where the story happened. When the professor realizes that the researcher was taking too long to come back, she starts to get concerned and decides to go inside too. This makes her visit her own ghosts, which was precisely the intention of the researcher.

Later, in the professor's apartment, she tries to explain. The real reason for her attitude had been information received by her husband that it was a trap by the Nazis to catch "traitors". If they helped the little girl they would have been caught. The researcher doubts it. So do we, the viewers. It would be too easy. In any case, the professor confesses that she has never been the same ever since. That is when the following dialog takes place:

Jewish researcher (Pj): What do you say to your students about how they ought to live?
Professor of ethics (Pe): I help them arrive at their own conclusions.
PJ: What do you mean?
Pe: The 'good'. That exists. I believe the good exists in all men. A situation may trigger the right or the wrong. That night [in which we didn't help you] it was not the 'right'.
PJ: And who judges what is right and what is wrong?
EP: That which is within you.

The first part of this dialogue is very true. An ethics professor does not teach anyone to be ethical. And if she claims to do that she is already in the wrong way. She will teach different possible approaches to ethics, the history of moral philosophy, different codes of ethics, propose moral dilemmas to be discussed. She instigates, nudges, bothers. The more she disturbs the students' inertia, the better. At the end, the best she can hope for is that their pupils leave her with a greater critical spirit.

Which brings us to the second point of the dialogue. Who judges? In fact, 'I' judge, that is inevitable. However, that is not enough, because, I, as a judge of myself, am not good enough. I commit my crime or misdemeanor and, if I do not get caught, I rationalize the guilt to the point of forgiveness. Forgiveness of myself, that is. After all, Dostoyevsky's Raskolnikov is just a fictional character. Nobody gets consumed by guilt until becoming sick and then gives himself up. As well as "no one is just for one's own will, but is forced. Whenever one has an opportunity, one commits an injustice, but tries and is careful not to be caught", as Glauco reminds us, after telling the story about the ring of Gyges in the second book of *The Republic* (Plato 2001, p. 360 c–d).

As for the 'right'. Does it even exist? Is it absolute or relative? And relative to what? To my will? To yours? To God's? In continuation of the conversation between the researcher and the professor, at first the professor suggests that the one who decides is God, but then she turns to a Karamazovian solution – man is free to choose. Or not. Since there is no 'right', this is relative and, as such, a collective construction. Man is not free to choose. Someone else chooses for him. The Heideggerian *Das Man* (the "they") dictates what is 'right' and what is 'wrong'. And if one tries to get rid of it, to be authentic, he is crushed. Of course, to be authentic is not necessarily a virtue. Hitler was authentic. And so was Stalin.

In addition, the mediocrity dictated by *Das Man* ('mediocrity' in the Aristotelian sense of the word) is also not necessarily a vice. If we assume that ethics is what makes a harmonic coexistence among individuals possible, every norm that contributes to it will be welcomed. And if guilt (my guilt) is not enough to give support to this, we always have indignation (of the other).

After seeing this movie, the following questions are good discussion's initiators:
What (really) is ethics?
What does a professor of ethics teach?
Is the professor of ethics more ethical than the rest of the faculty?
Is a student who gets an 'A' in an ethics test more ethical than the ones who got 'Bs' or 'Cs'?

10.3 Kant vs. Utilitarianism

The Chilean writer Ariel Dorfman drew inspiration from Schubert to write his play *Death and the Maiden*, which originated the movie directed by Roman Polanski, with the same title, starring Sigourney Weaver, Ben Kingsley and Stuart Wilson.

As the film is fairly faithful to the play, we will consider Dorfman as the author of the concepts that we will discuss, not Polansky.

Paulina Salas, a medical student and the girlfriend of Gerardo Escobar, was hijacked by government militia during a dictatorship. The reason for that was the fact that Paulina and Gerardo edited a student's subversive newspaper. Dorfman (1992) was careful not to place his story specifically in Chile, but in "any country that elected a democratic Government soon after a long period of dictatorship". They torture her to give the name of her companions, particularly that of Gerardo. At some point, they bring a doctor whose job was to assess to what extent the torturers could get without killing their victims. Blindfolded the whole time, Paulina, who never denounced any companion and endured horrors because of this, could only identify her abuser by the smell, voice, touch and taste for Schubert. Whenever he raped her, the doctor played the song *Death and the Maiden*. Ironically, Schubert had been, until then, one of the Paulina's favorites.

Several years later, Paulina is married to Gerardo. She did not graduate; therefore, she did not become a doctor. Gerardo, for his time, became an eminent lawyer, who had just been appointed Chairman of a Commission that would investigate the tortures. But there was a problem. This Commission would handle the cases that ended up in death, yet without revealing the names of the torturers. Paulina doesn't accept it, she considers it a farce.

As fate would have it, on a rainy night, after having the car stuck on the road back home (far away from the city), Gerardo was aided by a helpful doctor, Dr. Roberto Miranda. Paulina recognizes his voice. It was him, the same taste for Schubert (he even had a tape with Schubert's *The Death and the Maiden* in the car), the same habit of quoting Nietzsche, the same laugh, the same smell. She waits until her husband and the doctor sleep, then she ties the doctor to a chair and starts the trial, to the dismay of Gerardo. That's when the movie really takes off. The trial of the (supposed) torturer by his victim, having one of the best lawyers in the country to defend him, who just so happened to be the victim's husband. Dorfmann (and Polansky) can make the viewer feel exactly like Gerardo, switching from moments of doubt to absolute certainty – that's got to be him! – until the end.

The declared intent of the Chilean writer was political. Being himself exiled (though not tortured), Ariel Dorfman wants to touch on sensitive points of the recent political history of Chile and other Latin American countries that have gone through similar experiences. He had enormous difficulty, because the play was written and staged shortly after the beginning of democratization, which made some people criticize him, fearing the wrath of the military, still in control of the situation.

But there's a more interesting philosophical nuance. *Death and the maiden* reverses the roles usually assigned to Kantians and their opponents, the utilitarians. Usually, Kantians are regarded as excessively scrupulous, and utilitarians as cold calculators. The prototype of this in art (literature, more specifically) is the 'schizophrenic' Dostoyevsky's Raskolnikov and his struggle with his two selves, a utilitarian, which kills the old pawnbroker and elaborates a whole theory about "the extraordinary man" who does not hesitate to step on human heads to make the world go round, and another one, Kantian, which makes him literally sick of guilt for murdering the innocent sister in the process.

Paulina, the one who was tortured, wants to do justice with her own hands by killing her torturer. Gerardo, her husband, despite the chance of her being right, doesn't agree. He believes that if they do that they will jeopardize the process of democratization. Pauline's attitude could be interpreted simplistically as wanting only revenge. But she may also be considered a Kantian. Let's do the categorical imperative test: could we transform into universal law to punish with death a person capable of a heinous crime such as repeated torture and rape, ensuring in the process that the victim does not die so as to be kept being tortured indefinitely?

What about Gerardo? As a brilliant lawyer, he was determined to bring the accused to justice. But what justice? One may say the justice allowed by the limits imposed by the circumstances. Let's try a utilitarian calculation: is that attitude likely to be as useful as possible to the greatest number of people for the most prolonged time? If you push too hard you run the risk of provoking the military and everything goes back the way it was. But the limits imposed by the circumstances are exasperatingly small. Let's try and calculate again: what would be best, to act with more firmness and risk losing what has been achieved, or to follow the rules and keep the achievements? Or, as Dorfman (1992, p. 83) says: "it is legitimate to sacrifice the truth to ensure peace?"

Note that the conscientious here is Gerardo, the utilitarian, not Pauline, the Kantian. The viewer establishes an empathy with both throughout the movie but ends up agreeing with Gerardo.

10.4 An Enemy of the People – Ethics and Public Health

The brilliant Norwegian playwright Henrik Ibsen (1828–1906) wrote the play *The Enemy of the People* hastily and as a kind of response to the harsh reaction of the public to his previous play, *Ghosts* (McFarlane 1988).

Dr. Stockman is a conscientious doctor, in charge of a health bath facility in a small coastal town in southern Norway. Its waters were believed to have healing properties. However, Dr. Stockman noticed the appearance of new diseases in its users, which made him ask for an examination of the waters. When the result of the analysis arrives it is not encouraging. The waters were, as he suspected, contaminated. What to do next is obvious for the good doctor – he should warn everyone, close the facility, and decontaminate it. Simple.

Or not. Although Thomas Stockman is able at first to convince the local press about his intentions, the same does not occur in relation to Peter Stockman, the Mayor (who happens to be his brother). The reaction of Peter is classic, when it comes to politicians. If Thomas took out his plan, the city would be damaged. Its main source of income could be lost perhaps forever, because credibility is everything for something that involves health, and it is hard to win back.

The feeling we have when we read or watch the play is that the doctor is the only one genuinely worried about the people, not only of his own city, but also the tourists who come to the health bath. That is, he would be the only ethical character in the plot. This is clearly the intent of Ibsen. There is virtually not a single good thing left, not even a belief in democracy, because it is exactly its typical tricks which are used just to silence Dr. Stockman.

However, there are other possible readings. Terrence McConnell (2010) in his article *Moral Combat in An Enemy of the People: Public Health versus Private Interests* makes an interesting analysis from different perspectives. According to McConnell, it is possible to adopt the point of view of the Mayor and of the journalists, as well as the wife of the doctor, all contrary to his intentions, all with their own moral reasons.

The Mayor would not be all that wrong to consider that the damage to the city would not only be big, but unrecoverable. He would not be necessarily wrong by planning to make a slow, progressive and quiet decontamination. Of course, in this case, one can always imagine that the politician will do what politicians typically do – to not keep his promise. After all, those looking for the waters of the Spa are already sick. Furthermore, are the results of the tests of contamination reliable? – questions the Mayor.

Such arguments are able to convince with some ease the journalists, who, in addition to that, would lose their source of publicity. However, the doctor's wife is not convinced, but scared:

> But, Thomas, my dear, your brother has a lot of power on his side.
> _ Yes, but I am on the right side!
> _ Right! Yes, of course. What's the point of being on the right side without power? (Ibsen 1988, p. 41)

The concern of the wife is the family. Although Thomas was right, nothing would help fight all the forces against him. He would end up harming his own family (which is what actually happens).

Stubborn as all legitimate Kantians, Dr. Stockman won't give up. Not having the press by his side, he convenes a meeting to warn everyone and propose the immediate interruption of the functioning of the health bath facility. And that's when we see how the gears of democracy can serve paradoxically anti-democratic purposes. The citizens vote to forbid him to talk, and when he exasperates, they "elect him" enemy of the people, and he exasperates even more, which only makes things worse, causing those who were not familiar with the subject to think he was crazy or drunk.

In fact, although we are able to do an exercise of trying to see things through the different perspectives of each group, as the play walks to the end, the masks fall out completely. The allegedly valid arguments of doctor's opponents grow pale in the shadow of the real motivations of the Mayor and of the editor of the newspaper. The latter does not even try to disguise:

> Hovstad – And you mustn't forget that this business about the shares can be presented in two very different ways.
> Dr. Stockman- Yes, and you're just the man to do it. If I don't come to the aid of the Herald, then you'll take a pretty poor view of things. The hunt will be up. I dare say... You'll be after my blood… You'll be on me like a dog on to a hare.
> Hovstad – That's the law of nature. Every animal must fight for survival. (Ibsen 1988, p. 101)

Thomas Stockman, then, is a moral hero. If we consider his stubbornness in making public a scientific evidence damaging to the health of the users of the health bath facility, yes. However, it is worth asking: what about before that? Was there any solid evidence that the waters were good for the treatment of any disease? If they weren't contaminated and he hadn't ordered the tests, would he be an accomplice to a scam? That is a pertinent question even if we consider the time when the play was written (1882). What is neither good nor bad is allowed? Not in public health. Innocuous treatments can cause as much harm as harmful treatments. The money spent, the hopes, the deception. It is also immoral.

10.5 A Clockwork Orange and Free Will

In a way, although the movie *A Clockwork Orange* was based on the novel by Anthony Burgess, one could say that its authorship belongs to the director Stanley Kubrick. For a curious reason: Kubrick based his movie on the American, not the English edition of the book. Which means that he eliminated the last chapter, making his ending markedly more pessimistic than the book's (Morrison 2000, p. xvii–xviii).

Alex (Malcolm McDowell in his best role ever) is a teenage punk leader of his own gang. He spends his days practicing what he and his fellows liked most – ultra-violence, especially against other gangs. They enjoy beating drunks and beggars, they steal, rob, and rape. But Alex is different. He is fond of music. Not just any music, but mainly classical music. The Ninth Symphony of Beethoven (Ludwig van) gives him special pleasure. His associates do not have the same tastes. One of them dares to make fun of a lady who was singing the "Ode to Joy" (Fourth movement of the Ninth) in a bar, which forces Alex to punish him. It is the beginning of the end of the gang. The others try a reaction, which is vehemently rejected by their leader.

One day the opportunity for revenge appears and Alex is betrayed by his companions. Arrested, he spends 2 years in a maximum-security jail, becoming a friend of the priest. When the Minister of the Interior offers the opportunity to participate in a new program – the Ludovico method (a play on the first name of Beethoven) – Alex grabs it without hesitation.

The "treatment" consists of making the "patient" watch violent films, held in a chair with a blepharostat that prevents him to close his eyes. Such sessions are preceded by drug applications to enhance the desired effect, so that the patient gets really sick whenever he watches such scenes. There is, however, one detail particularly cruel: the background music is Alex's favorite, the Beethoven's Ninth.

When Alex is considered "cured" a demonstration is staged to show the success of the treatment for the Minister of the Interior. After the grotesque display of subservience (to use the words of the priest of the prison), in which Alex is no longer the old Alex, the priest protests:

> _Choice. He has no real choice, has he? Self-interest, fear of physical pain, drove him to that grotesque act of self-abasement. Its insincerity was clearly to be seen. He ceases to be wrongdoer. Her ceases also to be a creature capable of moral choices. (Burgess 2000, p. 94)

Obviously, both the doctor as well as the Minister of Interior debunk such arguments with pragmatism. What matters is that he no longer commits crimes. Besides, the prisons were already very crowded, and the conventional way of dealing with the problem was fruitless and not profitable.

10.6 Free Will

Are we really free? The relevance of this question is closely linked to another essential problem: are we responsible for our actions? We can only be punished if what we choose to do is, in fact, our choice, not written in the stars or in our genes. Punished?

Impossible not to be reminded here of Nietzsche:

> We don't have today any kind of sympathy with the idea of "free will": we know very well what this is all about – the ability of worst theological reputation to make humanity "responsible", in the manner of theologians, which means: to make humanity dependent on the theologians. (...) Every time one talks about responsibilities, it is usually the instinct to punish and to judge who is in action. (Nietzsche 2008, p. 56–57).

In this regard, it is symptomatic that the only voice raised against the conditioning of the young Alex to do only the good (or, better, stop doing evil) at the expense of the loss of his free will was the priest. Nietzsche would say: "Didn't I warn you?!".

But the question remains: are we really free? Does it matter to be really free? Outside the realm of Philosophy that question makes no sense. The common man doesn't even consider the possibility of it. Of course, we are free, he will say. The common man, however, believes in God. Well, if God is omniscient, omnipotent, has always existed and always will exist, God has seen this movie. Not *A Clockwork Orange*, but the film of our lives, from the beginning to the end. But if 'Someone' already knows the ending of the movie of my life, what I will do is already decided. Ergo, I'm already bad or good, I have no choice. Ergo, I can't be held responsible for my actions, right?

Wrong, the German philosopher Leibniz would answer. His Theodicy is an attempt to try to accommodate, in the same small and tight little box, the free will with the omniscience and the divine goodness. Pangloss, I mean, Leibniz, tries his best to convince us that our little doomed world is the best of all possible worlds, despite countless evidence to the contrary.

If a Theodicy, however elegant, is incompetent to overthrow the determinism inevitably linked to the idea of God and his attributes, you'd think the light at the end of the tunnel was on the opposite side, the biologist atheism. Negative. What is more structural than the genes? We are born with our programs already designed. Diseases, unfortunate decisions that we make, or will make, anyway, nothing is our fault, right? Who knows. Determined or not, what matters is that we don't know. It is the epistemic uncertainties that count, as Daniel Dennett (2003) proposes. We have to play the game (of life) with the available rules. And these require our ignorance. We are, therefore, responsible for our acts. Written in the stars (or proteins) or not.

10.7 The Trial of God

Pogrom is a word of Russian origin used almost as a synonym for the massacre of Jews. One of these pogroms happened in the small town of Shamgorod, in 1649, in Ukraine. Only two Jews survived, Berish, the innkeeper, and his daughter Hannah, who was getting married. To add to the cruelty of their act, the aggressors made the father watch the rape of his daughter by several of them. Shortly after that, it was the time of Purim, a festive date for the Jews, during which they celebrated victory over their enemies, as told in the book of Esther. It is common to represent small plays and songs alluding to the theme. Three minstrels are in the tavern of Berish and do their part in exchange for food and drink. Understandably, Berish wasn't in the mood. He was concerned only with his sweetheart Hannah, and had the support of his Christian maid Mary.

After the minstrels Mendel, Avrémel and Yanked insist a lot, so Berish concedes. On one condition, that it would have to be his choice. They would represent the trial of God. Berish, full of hatred for him, would be the prosecutor. Mendel, the wisest of the three, would be the judge. All ready. Or almost. A part was missing. Who would do the lawyer? Nobody wanted, each one with their own reasons. Because of the impasse, they nearly canceled the trial, until a stranger appears.

Sam, as the stranger wanted to be called, went on to show how good a lawyer he could be. He wouldn't let himself be shot down by any of the Berish's arguments, however convincing they were. For example, when Berish accused the defendant of hostility, cruelty and indifference to let his people be massacred, Sam is not satisfied and asks for evidence. Berish delivers-the Pogrom that occurred recently, and of which he was a witness. Sam, for his time, argues that the pain (of Hannah and Berish) did not constitute legal evidence. And that men were killed by men, where was God in all this? On the indignation of Berish, Sam explains: Berish saw God in the aggressors, while the right thing would be to see him among the victims.

The trial runs until the end with neither side giving in. Then, the priest of the town comes to warn them: an angry mob formed again, they needed to escape. But it was too late. When the curtain closes, the last revelation: Sam, the dedicated defender of God, was none other than the devil.

The above is a summary of the play *The Trial of God*. Its author is the winner of the 1986 Nobel Peace Prize, Elie Wiesel, Romanian Jew born in 1928 and survivor of Nazi concentration camps. There is a French TV movie called *Le procès de Shamgorod* which was based on the play.

The idea of proposing a trial of God is both courageous and unusual. Few would dare, especially with an unhappy ending, with God being sentenced. Among the few who earned this right certainly are Holocaust survivors like Wiesel. What they suffered and witnessed is "legal evidence" enough, despite what His lawyer says. And the fact that this lawyer is none other than the devil is absolutely original. Particularly considering that he tried to deconstruct the wrath of Berish.

But what really intrigues is taking the trouble to make a trial of God. After seeing his family and friends dead and his daughter raped again and again, it is not surprising that Berish is mad at God, but that he still believes that He exists. The real trial didn't happen. And before anyone says it's because the play was set in 1649, and that's why something like that couldn't have crossed the mind of the characters, it's good to remember that Leibniz, the author of Theodicy, was born in 1646.

The trial could have happened in a different way. If, for example, Leibniz were to try to convince Berish "with mathematical evidence" of the existence of God, the former would refute the latter arguments with facts. He would ask the philosopher to justify the cruelty of pogroms. Wiesel himself would probably ask for a justification of concentration camps. Gottfried Leibniz (2007) could try and answer: "Nor could I ever accept the view of some recent philosophers who have the nerve to maintain that God's creation is not utterly perfect, and that he could have acted much better. This opinion, it seems to me, has consequences that are completely contrary to the glory of God. Just as a lesser evil contains an element of good, so a lesser good contains an element of evil. To act with fewer perfections than one could have done is to act imperfectly; showing an architect that he could have done his work better is finding fault with it. Furthermore, this opinion goes against holy scripture's assurance of the goodness of God's works. That goodness can't consist simply in the fact that the works could have been worse; and here is why·. Whatever God's work was like, it would always have been good in comparison with some possibilities, because there is no limit to how bad things could be". One can only imagine Berish strangling Leibniz after hearing such a cool discourse, detached from reality.

10.8 Final Considerations

At this point we can submit the films suggested here to the screen of the interesting proposal of Jan Solbakk (2015). For him, although all traditional forms of (bio) ethics teaching are valid, cinema (as well as theater) is the best of them, since it is capable of instigating dialogue in the Socratic-Platonic molds, besides being tragedies, contemplating, therefore, the Aristotelian approach, for whom the latter have the capacity to move us. In other words, they are films that lead us to a catharsis, both as purification (platonic) and as *hamartia* (Aristotelian). The films suggested here, as well as several others that can be used, have the characteristic of putting the viewer against the wall. It is worth noting that 'spectator' means both the student and the teacher, which again brings us back to the Socratic-Platonic dialogue. There is not just one correct answer. The definitive solution is not in the teacher's possession for the simple reason that it does not exist. According to Solbakk, it is the dialogue itself that ends up having the therapeutic function of catharsis, in the sense of a purification of our self-indulgence and moral indifference. The result is (or should be) a healthy angst.

10.9 Some Other Suggestions

One of the biggest advantages of movies as instruments for discussions in class, even when they are versions of another literary medium such as plays or novels, is their ability to touch on different global moral problems. The examples given here evidently do not exhaust the possibilities offered by this tool. *The Constant Gardener* (2005), based on a novel by John le Carré and directed by Fernando Meirelles, is great to discuss the role of Big Pharma and the ethics of research in humans. *Miss Evers' Boys* (1997), based on the play by David Feldshuh and directed by Joseph Sargent, narrates the story of the infamous Tuskegee study, therefore, it is perfect to discuss racism and the ethics of research in humans. *Crimes and Misdemeanors* (1989), written and directed by Woody Allen, shows wonderfully how unreliable a moral judge our own conscience is. And there certainly are many more good examples. We encourage the teachers and the students to try and find the ones that touch them the most.

References

Burgess, A. 2000. *A Clockwork Orange*. London: Penguin Books.

Dennett, D. 2003. *Freedom Evolves*. New York: Penguin Books.

Dorfman, A. 1992. *A morte e a donzela*. Rio de Janeiro: Paz e Terra.

Ibsen, H. 1988. *An Enemy of the People. The Wild Duck. Rosmersholm*. Oxford/New York: Oxford University Press.

Leibniz, G.W. 2007. *Discourse on Metaphysics*, 2. Available at http://www.earlymoderntexts.com/assets/pdfs/leibniz1686d.pdf.
McConnell, T. 2010. Moral combat in an enemy of the people: Public health versus private interests. *Public Health Ethics* 3 (1): 80–86.
McFarlane, J. 1988. Introduction. In *An Enemy of the People. The Wild Duck. Rosmersholm*, ed. Henrik Ibsen, ix. New York: Oxford University Press.
Morrison, B. 2000. Introduction. In *A Clockwork Orange*, ed. A. Burgess, vii–xxiv. New York: Penguin Books.
Nietzsche, F.W. 2008. *Crepúsculo dos ídolos. Ou como filosofar a marteladas*. 2ª. ed., 56–67 (Translated from German to Portuguese by Antônio Carlos Braga). São Paulo: Editora Escala.
Platão. 2001. *A República*. 9a. Ed., 58 (Translated from the Greek to Portuguese by Maria Helena da Rocha Pereira). Lisboa: Fundação Calouste Gulbekian, (361.b).
Solbakk, J.H. 2015, September. Movements and movies in bioethics: The use of theatre and cinema in teaching bioethics. In *Bioethics Education in a Global Perspective*, ed. Henk ten Have, 203–221. Dordrecht: Springer Publishers. doi: https://doi.org/10.1007/978-94-017-9232-5_16.

Chapter 11
The Role of Case Studies in Global Ethics Education

Willem A. Hoffmann

Abstract Case studies are widely used in ethics education. This chapter addresses four main areas: (1) Definition of "case study"; (2) Educational value of case studies; (3) Case study content; and (4) Case study scheduling. Various educational aims can be pursued through case studies: facilitation of self-reflection; understanding of ethics principles; appreciation of real-life ethical challenges; and critical reasoning skills. Case study content is guided by educational factors and case study content sources. Five educational factors are considered: (1) Congruence with course learning objectives and unit themes; (2) Congruence with educational objectives; (3) Content fidelity; (4) Contextual focus; and (5) Level of complexity. Various case study content sources can be utilised, namely casebooks, audio-visual material, news reports, formal literature and self-developed material. Some practical considerations are applicable when selecting case study format in a specific context. The first is the case study length, followed by the students' learning style preferences. The third is case study frequency during a course, while the fourth consideration focuses on using an educator-facilitated approach or a student-based self-guided approach. Case studies can be scheduled at the start of a session where it raises awareness and is useful for educator situation analysis. During a session it illustrates concepts, stimulates reflection and develops analysis skills. At the end of a session it is beneficial for concept integration and formative assessment, while between sessions it is applicable for formative assessment, self-reflection and analysis skill development. At the end of the course it is suitable for summative assessment.

W. A. Hoffmann (✉)
Tshwane University of Technology, Pretoria, South Africa
e-mail: HoffmannWA@tut.ac.za

H. ten Have (ed.), *Global Education in Bioethics*, Advancing Global Bioethics 10,
https://doi.org/10.1007/978-3-319-78984-2_11

11.1 Introduction

Teaching bioethics is a challenging endeavour for ethics educators, especially given the notion that "… ethics cannot be taught … but rather can be learnt through observational and educational learning, and through experiential participation" (Ssebunnya 2013, p. 49). The implication is that the primary emphasis in global ethics education must be underpinned by a transfer of the educational focus away from "teaching" to "learning" (Potter 2015).

Traditionally, didactic lectures have been the most common primary teaching activity in the majority of educational contexts around the world (Potter 2015). Educators who follow this mostly one-directional approach tend to almost exclusively focus on theoretical knowledge transfer of ethical concepts, principles and theories from the educator to the students (Stolper et al. 2016). In recent years many educators have increasingly started to make use of active and interactive learning approaches such as case studies, role plays, small group discussions and self-reflection activities (Potter 2015; Stolper et al. 2016; Ten Have 2016; Vaswani 2016). Active learning focuses on activities that engage students in doing, experiencing and reflecting on the course material, especially with ideas and principles that can be applied to practical examples in real-life contexts (Ekberg 2016; Potter 2015; Vaswani 2016). Case-based learning is an interactive, student-centered educational approach that specifically facilitates productive learning-oriented discussions, dialogues and reflections (Wilt 2016). Other important benefits of this approach are its potential to generate alternative viewpoints and responses to case-based dilemmas, as well as its potential to engage in peer-based learning (Ekmekçi and Arda 2016). A recent study by Ekberg (2016) in the United Kingdom found that university students who are enrolled in bioscience courses prefer a teaching strategy consisting of lectures, case studies and seminars. The study also found that students prefer to be assessed by means of case studies that require a written critical analysis. Such a multimode approach to ethics education is ideal to apply ethical theories and principles to real-life scenarios and to develop and assess critical reasoning skills (Ekberg 2016; Wilt 2016).

Case studies can be effectively used to teach factual knowledge, to apply moral theories to real-life contexts, to develop ethical analysis and reasoning skills, and to develop awareness and sensitivity to ethical issues. Also, it is useful to foster dialogue, reflection, deliberation and experiential learning, especially when used in combination with small group discussions and other non-traditional approaches such as moral games (Gracia 2016; Ekberg 2016; Vaswani 2016; Wilt 2016). The ability to effectively engage in dialogues fosters critical self-reflection and facilitates the development of listening skills, as well as the skills to ask appropriate questions rather than merely focusing on convincing others of one's own viewpoints (Stolper et al. 2016; Wilt 2016). As such, global ethics education activities should specifically place a high premium on the development and mastering of deliberation skills. The implication is that the goal of these activities should never be to provide recipes or algorithms to arrive at the "correct" answer to ethical challenges, but to effectively use deliberation skills to arrive at "wise" and reasoned decisions and positions (Gracia 2016).

Effective global ethics educators should have a high-level awareness and application of educational principles and methods to create optimal contexts for students to internalise ethical skills. Case studies are widely used instruments in this regard. This chapter addresses the following main areas: (1) Definition of "case study" concept; (2) Educational value of case studies; (3) Case study content; (4) Format of case study presentations; and (5) Case studies scheduling.

11.2 Definition of "Case Study" Concept

A *case study* is "a documented study of a specific real-life situation or imagined scenario, used as a training tool … (in which) students are required to analyze the prescribed cases and (to) present their interpretations or solutions, supported by the line of reasoning employed and assumptions made" (BusinessDirectory 2016). Case studies often involve descriptions or representations of ethical dilemmas that require a need for compromise. However, the educational value of case studies should include more than theoretical reflections and analyses of various case scenarios. Case study analyses and presentations should also include realistic, action-based recommendations and solutions that can ideally facilitate the long-term transfer and application of ethical principles and concepts to future post-qualification professional contexts (Potter 2015).

11.3 Educational Value of Case Studies

It is important to recognise that case studies are *educational instruments* that are not an end in itself, but rather a way to facilitate effective active learning. The key question in this section is the following: *Why* are case studies used in ethics education contexts? In essence, case studies facilitate the following four educational outcomes: (1) Self-reflection on personal values and value positions; (2) Identification and understanding of ethics principles and concepts; (3) Appreciation of real-life ethics challenges in personal and professional contexts; and (4) Critical reasoning skills.

Case studies facilitate the process of *self-reflection* in which students develop skills to recognise and replace rigid dogmatic thinking styles with dynamic value-oriented thinking styles. Reflection also facilitates the development of skills to acknowledge and appreciate others' points of view. As such, reflection activities help students to actively engage with and to analyse the ethical complexities of case studies (Vaswani 2016; Wilt 2016). Case studies can also be used to facilitate the development of various *critical reasoning skill* aspects on a personal and group level (Stolper et al. 2016). One such aspect is the ability to analyse and evaluate a case's cultural and social contexts. Another aspect is that case study discussions facilitate the ability to question and use critical thinking in identifying values and principles (Vidal 2016).

11.4 Case Study Content

Two key areas are addressed in this section. The first area focuses on the educational factors that contribute to the content of case studies in order to facilitate effective learning educational contexts. The second area focuses on the most important sources of case studies.

11.4.1 Case Study Content – Educational Factors

A number of educational factors contribute to case study content. The first factor is that the case study content must be *congruent* with a specific course's overall *learning objectives and unit themes*. The learning objectives guide decisions regarding the appropriateness of the case material to best meet the students' needs and to achieve specific education outcomes (Wilt 2016). However, specific learning objectives should be identified and formulated for each case study to align it with the specific session's learning objectives and the overall course objectives (Ekmekçi and Arda 2016).

In a global ethics education context, the UNESCO Bioethics Core Curriculum (UNESCO 2008) and the Universal Declaration on Bioethics and Human Rights (UNESCO 2005) provide a framework for the identification of the most important unit themes that should ideally be addressed in a bioethics course. The case study content should then be linked, implicitly or explicitly, with one or more of these units. The following main themes are applicable: (1) Human dignity; (2) Benefit and Harm; (3) Autonomy; (4) Consent (including Capacity to Consent); (5) Vulnerability and Personal Integrity; (6) Privacy and Confidentiality; (7) Equality, justice and equity; (8) Non-discrimination and Non-stigmatisation; (9) Cultural diversity and Pluralism; (10) Solidary and Cooperation; (11) Social responsibility; (12) Sharing of benefits; (13) Protecting future generations; (14) Protection of the environment, the biosphere and biodiversity (UNESCO 2008).

A second factor that contributes to the appropriateness of case study content is its *congruence* with the *specific educational objectives* that the educator wants to achieve in the educational context. According to the revised Bloom's taxonomy (Krathwohl 2002) these objectives may vary from a focus to merely remember specific facts to the ability to demonstrate high-level reasoning skills. It describes six major objectives/levels that might individually or in combination be pursued by case studies. The specific educational objective/s of a case study will depend on its specific aim in a given context. The implication is that not one of these six objectives/

levels is in itself more or less appropriate or "better"; each objective has educational value to achieve a particular learning outcome in a particular context. The most basic taxonomy level focuses on the objective to *remember*, recall and recognise specific theoretical facts, concepts or procedures. The next level focuses on the objective to *understand*, comprehend, explain, classify or interpret the meaning of specific factual information, ideas or concepts. The third level focuses on the objective to *apply*, which is essentially the skill to implement and transfer theoretical knowledge and skills to real-life contexts (Wilt 2016). This is followed by the objective to *analyse*, to differentiate between closely related concepts, and to determine and identify the interactions, patterns and components of models, theories and real-life events. The fifth level of the taxonomy focuses on the objective to *evaluate*, to make judgements based on specific criteria, to justify a position and to offer critique on theories, models and others' arguments. The last, and arguably highest, level focuses on the objective to *create*, generate or produce new/novel patterns and structures. On this level, Krathwohl (2002) indicates that all five previous levels of learning are involved to some extent. When students are required to create case studies they are also per implication required to know the facts, to understand the meaning of concepts, to apply it to the relevant context, to analyse the interactions and components, to evaluate and critique different viewpoints, and finally to create a new case study that illustrates the integration of concepts and principles. On this level students can be required to formulate (create) case studies involving possible ethical concerns and dilemmas from personal contexts and personal experiences or contemporary ethical issues (Potter 2015).

The third educational factor that contributes to the appropriateness of case study content is *fidelity*. It refers to the level of believability or realism portrayed in the case study. Fidelity can vary from a low level (i.e. unrealistic case scenario) to a high level (i.e. authentic case scenario). Generally, a high level of fidelity facilitates the student's ability to suspend potential scepticism and to become significantly immersed in the case scenario and its ethical dilemma/s. This, in turn, creates an authentic learning experience in which the student applies and develops critical thinking and analysis skills (Wilt 2016; Kenny et al. 2015). Authentic case scenarios more accurately reflect real-life scenarios that students may encounter in their future or current professional contexts (Kenny et al. 2015; Potter 2015). Hypothetical cases are often developed in a deductive approach by using specific theoretical concepts, principles and values which are illustrated in a specific case scenario. The aim is then to specifically expose students to these concepts, principles and theories (Stolper et al. 2016; Wilt 2016). Hypothetical case scenarios are also useful to expose students to potential future dilemmas and challenges (Ekberg 2016). However, a significant limitation of hypothetical case studies is that the relevant principles, values and theoretical frameworks have been determined *before* the formulation of the actual case content, which is contrary to the order in which real-life case scenarios are encountered and analysed, namely firstly the case *followed* by an analysis and critical reflection on the applicable principles and values (Stolper et al. 2016).

A fourth educational factor that contributes to the appropriateness of case study content is its *contextual focus*. In a global ethics education context it is essential to use a combination of local, regional and international case scenarios. This raises students' awareness of global ethics issues while also allowing them to engage with ethical dilemmas they are familiar with or might encounter in their daily lives and professional contexts. On another level it is important to use case studies from all parts of the world and not only the "classic" well-known case studies from North American (e.g. Havasupai and Tuskegee cases) and Western European (e.g. Nazi concentration camps) contexts.

A fifth educational factor that contributes to the appropriateness of case study content is its level of *complexity* (Kenny et al. 2015). Generally, simple case studies that merely illustrate a specific concept will initially be used in a course. These will be followed by progressively more difficult/complex case studies that are often longer in length, which contains a combination of several ethics themes, and which requires from students to use application, analysis, evaluation and creation skills (Krathwohl 2002).

11.4.2 Case Study Content Sources

A variety of sources are utilised to obtain appropriate case study content in order to facilitate active learning. Also, the choice of case study source is determined by a number of practical considerations within each educational context.

11.4.2.1 Sources of Case Study Material

The most widely available sources of case study material include casebooks, audio-visual material, news reports, formal literature/documents and self-developed material.

Casebooks

Numerous casebooks in various formats, including electronic online formats, are available. Many of these casebooks contain real-life case studies specifically related to various aspects of medical ethics (Potter 2015), for example the two casebooks in UNESCO's Case Book Series that focus respectively on Human Dignity and Human Rights (UNESCO 2011a) and Benefits and Harms (UNESCO 2011b).

Audio-Visual Material

Videos present narrated factual information linked to visual images (Engel-Glatter et al. 2018). YouTube™ (www.youtube.com) is a rich source of case study material that covers almost any conceivable aspect of ethics and bioethics. Many of the video clips are relatively short in duration (< 15 min) which allows ample time for students to comprehensively discuss, analyse and evaluate its content in a class session (Ekmekçi and Arda 2016). A good Southern African example is the San Hoodia Project that deals specifically with benefit-sharing issues between the San people living in the Kalahari Desert and corporate entities; the focus is on the real-life context of the San's indigenous knowledge regarding the *Hoodia* plant's appetite suppressant properties (http://www.youtube.com/watch?v=1I295KAWwes). Videos can also be used to present arts-based material in which students are required to engage with and interpret the implicit and explicit ethics-related meanings of art objects. A good example is a video depicting the work of Ina Kooper, a Dutch painter, on human vulnerability (http://www.youtube.com/watch?v=mmgp_zm5rX0). Films are less frequently used than YouTube™ video clips in educational contexts, primarily due to its time duration (usually a minimum of 30 min, but more often between 60 min and 2 h). A distinct advantage of films is its complexity and extensive portrayal of authentic ethical dilemmas, together with rich character development of the role players (Ekmekçi and Arda 2016); see Chapter by Paranhos in this book for more on the use of cinema in global ethics education. Other types of visual material that can effectively be used as sources for case studies include cartoons, advertisements and various art objects/products (e.g. paintings, photographs, poems and songs). A good example of a content-rich cartoon depicts the internationally-known Emeritus Archbishop Desmond Tutu expressing his views on the right to choose a dignified death (Fig. 11.1). This cartoon raises several ethics issues such as morality, autonomy, integrity, justice, compassion and courage.

News Reports

News reports in printed or electronic format that deal with events-of-the day are particularly useful for case studies that address contemporary and real-life issues. Also, it provides rich sources of local, regional and international ethical dilemmas. Such case studies allow students to contribute additional facts and insights related to the case content that they might have encountered outside the educational context. Another advantage is that news reports are often presented in a form that can be used by educators without significant changes; which in turn allows educators to easily and regularly introduce new case study material into the course. One relatively recent regional example is a news report that appeared on News24 (29 Feb 2016), an online news site, regarding a developing refugee crisis in Malawi (http://www.news24.com/Africa/News/refugee-crisis-brewing-on-south-africas-doorstep-20160229). This news report addresses several ethical issues regarding human rights, human dignity, vulnerability, discrimination, stigmatisation and justice. Its regional significance for Malawi, Mozambique, Zimbabwe and South Africa addresses the complexity of real-life ethical dilemmas. This case study can easily be linked to global ethics issues regarding refugees, especially the current Syrian refugee crisis.

Fig. 11.1 Cartoon "Desmond Tutu – South Africa's Moral Compass". A formal license and permission was granted by the cartoonist, Jonathan Shapiro (Zaprock Productions) on 29 Nov 2016 to reproduce the copyright-protected cartoon (150825tt)

Formal Literature

Formal literature refers to academic works (e.g. monographs and theses), peer-reviewed journal articles and court records. The cases reported in these works are usually authentic, credible and based on verifiable facts. In addition, these cases are often extensively discussed by the authors with reference to other formal literature. A North American example is the scholarly journal article by Palys and Lowman (2012) on the Boston College Belfast Project case that addresses issues regarding court-ordered disclosure of confidential data, the protection of confidentiality in sensitive research, ethics vs law conflicts, and challenges to academic freedom.

Self-Developed Cases

Experienced ethics educators and ethics committee members often have extensive professional experience that they can rely on to develop fictitious case studies, to develop tailor-made case studies that are loosely based on known events or facts, or to adapt existing case scenarios to meet specific learning outcomes and needs (Potter 2015; Wilt 2016). The advantage of these case studies is that they are often developed to meet specific educational objectives for particular student groups. Such case studies are especially useful in the early stages of a course to illustrate selected ethics concepts/principles and to raise students' general ethics awareness (Stolper et al. 2016). One potential negative aspect of self-developed case studies is its potential lack of authenticity, fidelity and/or transferability to real-life contexts (Stolper et al. 2016).

11.4.2.2 Practical Considerations in Selecting an Optimal Case Study Format

The first practical presentation consideration is the *length* of the case study. This includes considerations regarding the length of text-based case studies, as well as the time duration of audio-visual-based material. Case studies may be as short as a single sentence or paragraph (text), single image (picture, cartoon, painting), or 1 min (video, narration, music). However, extensive and complex case studies can be several pages in length or even a complete book (text), around 1 h in the case of documentary videos, and even longer in the case of full-length films. The appropriate length of the case study is inter alia determined by the specific objective of the case study and the attention-span of the students. Shorter case studies are appropriate for ice breakers in order to raise awareness or to illustrate single concepts. Longer case studies are appropriate to illustrate the ethical complexity of real-life scenarios and to facilitate students' analysis, evaluation and creation skills. Another important factor to consider is students' attention-span. It seems that modern-day adolescents have a relatively short attention span in educational settings which requires educators to pay careful attention to the length/duration of the case study material. However, it is also important for educators to ensure variation and frequent transitions in teaching and learning activities during a session in order to re-stimulate students' attention (Engel-Glatter et al. 2018).

The second practical presentation consideration when selecting the appropriate case study format is the students' *learning styles* (learning preferences). Perlman (2008) has described a useful model consisting of four primary learning styles that can specifically be applied to ethics education contexts, namely: (i) kinaesthetic; (ii) reading/writing; (iii) aural; and (iv) visual learning styles. It is important to keep in mind that any particular student usually prefers one of the four learning styles above the others. However, such a student can potentially still use and appreciate the benefits of the other learning styles to a lesser or greater extent when so required. Similarly, different learning activities, such as a case study, tend to be presented in a way that appeal more or less to one or more of the learning styles. Case studies that rely on written text and essay-based analyses will appeal more to students with a reading/writing learning style, while a cartoon-based case study will appeal to students with a visual learning style. Role plays primarily target kinaesthetic learning styles. Video clips and films offer students a wider preference range as it appeals to both aural and visual learning styles, and sometimes even to a reading learning style if it includes on-screen text. A group discussion of a case study in the format of a text-based hand-out can appeal to students with a combination of aural and reading learning styles. Ideally, educators need to be aware of all the students' learning style preferences and ranges when deciding on case study content, format and learning activities (Perlman 2008).

The third practical presentation consideration in selecting the appropriate case study format is *frequency*, i.e. whether to use a specific case once during a course to achieve a particular learning outcome or to use the same case study repeatedly at various points throughout the course offering. The advantage of once-off use of case studies is that students are at each point confronted with a new case study to facilitate new areas of self-reflection, critical reasoning and ethics knowledge (Vaswani 2016; Wilt 2016). In the case of repeated use of the same case study at various points during a course the advantage is that students can deeply engage with the case content and can develop high-level skills as the course progresses. A disadvantage of once-off case studies is that each introduction of a new case study during the course requires that all students must first become familiar with the case study scenario and content before further educational outcomes can be pursued. This might require considerable time for extensive, complex or long case studies; time which could've been used for other educational activities. A disadvantage of repeatedly using the same case study at various points during a course is that students might become "too familiar" with the case study content and as a result become "blind" to the finer nuances of the case which might ultimately inhibit deep learning and skills development.

The fourth practical presentation consideration in selecting the appropriate case study format is whether to use an *educator-facilitated approach* or a *student-based self-guided approach*. In the educator-facilitated approach the educator is primarily responsible for the following: to create a context of discovery amongst the students; to focus the students' attention on the central point/s; to facilitate and stimulate group interaction and dialogue; and to manage disagreements and/or conflict between students (Horn 2013). In the self-guided approach, the *student* is primarily responsible to engage in a process of self-reflection and self-discovery. In both approaches the case study material should ideally include the following elements: clear instructions; the actual case information; some theoretical information or guidelines; probing/guiding questions; and source references. Verbal and/or written instructions should clearly indicate the specific educational objectives, expected learning outcomes and, especially in the early stages of the course, guidelines to interact with the case study. These guidelines might highlight some of the important and applicable ethical concepts, principles and theories. It can be provided prior to the actual case study as part of the general teaching and learning activities in a session. Alternatively, the guidelines can be provided as an integral part of the actual case study material in order to provide students with sound theoretical underpinnings during the case study discussion. Another possibility is to only provide the guidelines as part of formative assessment activities during the feedback phase at the end of the session. Apart from guidelines, one or more probing/guiding questions may be provided to facilitate student learning. The use of appropriate and well-formulated open-ended questions can then facilitate critical self-reflection and group discussions (Wilt 2016). In the early parts of a course one might consider providing a number of directed and closed-ended questions, while increasingly more open-ended questions should be provided towards the end of the course.

Ultimately, a single evaluation-focused question may be posed in a summative assessment context, for example "Critically evaluate the position of the various role players with regards to the ethics dilemma in the case study?" Lastly, it is important to provide full references, including copyright permissions where applicable, to all the sources used and/or cited in the case study. It includes sources that might provide useful information and facilitate further learning regarding specific aspects of the case study. It is important to note that interactive, web-based ethics education offerings can potentially follow either an educator-facilitated approach or a student-based self-guided approach. The advantage of an interactive web-based offering is that it allows for synchronous and asynchronous educator-student and student-student interactions (Alessi and Trollip 2001; Ekmekçi and Arda 2016).

11.5 Case Studies Scheduling

Wise and informed scheduling of case studies in a course can significantly enhance its effectiveness to reach specific learning outcomes (Wilt 2016). Various educational objectives and learning outcomes can be attained by scheduling case studies at specific points during a class session and at specific points across the duration of the course (Table 11.1). Educators should consider the following key question when considering the inclusion of a case study in a session or during a course: "Why do I want to use this particular case study at this specific moment in the session/course?" Only valid educational reasons should be duly considered, while case studies should never be included or scheduled for entertainment or sensation-value purposes.

11.5.1 Start of a Session

At the start of a class session a short case study is often used as an ice-breaker to introduce students to a specific ethics concept. A brief reference to a current ethics dilemma can be used to engage students in the session's theme or to raise their

Table 11.1 Educational objectives for scheduling of case studies at various points in a course

Start of a session	During a session	End of a session	At home between sessions	End of the course
Ice-breaker and/or moral game	Illustrate concepts	Concept integration	Formative assessment (analysis, create)	Summative assessment (test/exam, create case study, write an essay)
Raise awareness	Stimulate reflection	Stimulate reflection	Self-reflection	
Presenter – Situation analysis	Develop analysis skills (individual and group)	Formative assessment (verbal)	Develop analysis skills (individual)	

general ethics awareness in order to set the scene for the rest of the session. Alternatively, a case study can be incorporated into a moral game at the start of a class session to raise ethical awareness in a fun context (Macer 2008). A different application of case studies at the start of a session is for the educator to conduct a covert situation analysis of the students' level of ethics awareness and knowledge. This information can then be effectively used by the educator to decide on appropriate educational strategies for the rest of the session.

11.5.2 During a Session

Case studies can be used during the main body of a session to reach various educational objectives. The first objective is to illustrate specific concepts and principles to the students. Usually these case studies will be relatively short and simple to allow students to easily grasp the intended knowledge and/or to comprehend the focus concept and/or principle. A second objective is to facilitate self-reflection regarding one's personal opinion of specific ethical issues, while a third objective is to apply ethics concepts and principles in the students' personal and current/future professional real-life contexts (Carlin et al. 2011; Ekberg 2016; Wilt 2016). Longer and more complex case studies can be used during a session to achieve higher levels of Bloom's taxonomy (Krathwohl 2002), namely to facilitate the development of deliberation and analysis skills. The students, either individually or in groups, will then be required to analyse the case study (i.e. identification of the ethical dilemmas, ethics principles, professional codes of conduct and legal considerations), to write a report and to make a class presentation (Ekmekçi and Arda 2016; Potter 2015). An advantage of group-based case study discussions is its ability to highlight and explicate the range of values held by students, including the range of relative importance attached to various moral values and ethical principles (Horn 2013). In becoming aware of others' insights, opinions, values and emotions the group members can develop skills to carefully listen to others, to appreciate and respect the views of others, and to actively engage with potentially conflicting inputs and arguments. Such group discussions effectively facilitate the learning of context-based concepts, principles and theories, while it also raises individuals' awareness and respect for alternative perspectives (Potter 2015; Schröder-Bäck et al. 2014).

11.5.3 End of a Session

Towards the end of a session case studies are useful to facilitate concept integration. New concepts and principles that were introduced and discussed earlier in the session can be effectively integrated with concepts and principles from previous sessions. At this point case studies can also facilitate self-reflection, specifically regarding one's personal opinions on specific ethical issues, and it can result in the application of ethics concepts and principles to the students' personal and professional real-life contexts (Carlin et al. 2011; Ekberg 2016). The process of reflection helps students to identify and appreciate the multi-layered nature of ethical complexities (Vaswani 2016; Wilt 2016). Lastly, case studies are ideally suited as tools for formative assessment during a session; i.e. to monitor and support progression in the learning process (Hoffmann and Nortjé 2016; Potter 2015; Wilt 2016). Constructive qualitative feedback, whether written or verbal, should ideally be provided in a supportive context that allows the students to further explore and reflect on their emotions, thoughts, insight and reactions (Taras 2009; Wilt 2016).

11.5.4 At Home Between Sessions

Case studies facilitate the development of students' analysis and creation skills (Krathwohl 2002) when used for formative assessment activities between sessions, i.e. at home during the course offering (Hoffmann and Nortjé 2016). The formative assessment activities should be clearly guided by specific learning outcomes and instructions. These instructions might also include open-ended guiding questions. In this regard, case studies are again useful to monitor and support progression in the learning process through various forms of qualitative feedback, in particular written and/or verbal feedback on submitted case study reports (Taras 2009). Case studies can also be used to facilitate personal self-reflection between sessions, especially when students are required to individually analyse and evaluate authentic context-relevant case studies by means of open-ended questions that probe their personal opinions and viewpoints (Ekberg 2016; Wilt 2016). A last application of case studies between sessions is to facilitate the development of *individual student-centred* analysis skills which addresses a different aspect of analysis skill development than the use of case studies during a session to facilitate *group-focused* analysis skills.

11.5.5 End of the Course

A generally accepted education principle is that valid evidence of student learning needs to be appropriately collected during and at the end of all programme offerings (Ten Have 2016). As such, case studies can be effectively used at the end of a course for *summative assessment* purposes (Hoffmann and Nortjé 2016; Wilt 2016). The evaluative function of summative assessment is to determine and/or make a judgment on how much of the course content a student has mastered; it pursues a performance evaluation objective (Taras 2009). Case studies can be used in written or oral tests/exams and a rubric can be employed to quantitatively score the appropriate use of reasoned arguments and responses to one or more questions about the case (Carlin et al. 2011). Alternatively, students might be required to submit a newly created case study grounded in contemporary real-life ethical dilemmas for summative assessment purposes (Carlin et al. 2011; Engel-Glatter et al. 2018; Potter 2015).

11.6 Conclusion

Well-designed, context-based and context-relevant case studies can be effectively employed in educational settings to facilitate self-reflection, bioethics knowledge, ethical reasoning skills and applications in real-life personal and professional contexts. Case study content is guided by various educational factors and the available case study content sources. A variety of practical considerations are applicable when selecting the appropriate case study format for a specific educational context, while case studies can be effectively scheduled throughout the duration of a course.

References

Alessi, S.M., and S.R. Trollip. 2001. *Multimedia for Learning: Methods and Development*. 3rd ed. Boston: Allyn and Bacon.

BusinessDirectory. 2016. What Is a Case Study? Definition and Meaning. http://www.businessdictionary.com/definition/case-study.html. Accessed 5 Nov 2016.

Carlin, N., C. Rozmus, J. Spike, I. Willcockson, W. Seifert, C. Chappell, P. Hsieh, T. Cole, C. Flaitz, J. Engebretson, R. Lunstroth, C. Amos, and B. Boutwell. 2011. The health professional ethics rubric: Practical assessment in ethics education for health professional schools. *Journal of Academic Ethics* 9: 277–290.

Ekberg, M.E. 2016. Exploring the design, delivery, and content of a 'bioethics for biosciences' module: An empirical study. *Journal of Academic Ethics* 14: 103–114.

Ekmekçi, P.E., and B. Arda. 2016. Education: Methods. In *Encyclopedia of Global Bioethics*, ed. H.A.M.J. ten Have, 1052–1059. Springer: Dordrecht.

Engel-Glatter, S., L.Y. Cabrera, Y. Marzouki, and B.S. Elger. 2018. Teaching bioethics to a large number of biology and pharma students: Lessons learned. *Ethics & Behavior* 28(1): 70–90.

Gracia, D. 2016. The mission of ethics teaching for the future. *International Journal of Ethics Education* 1: 7–13.

Hoffmann, W.A., and N. Nortjé. 2016. Education: Assessment. In *Encyclopedia of Global Bioethics*, ed. H.A.M.J. ten Have, 1012–1031. Springer: Dordrecht.

Horn, L. 2013. Promoting responsible research conduct in a developing world context. *South African Journal of Bioethics and Law* 6 (1): 21–24.

Kenny, B., M. Lincoln, and F. Killian. 2015. Ethics cases: Do they elicit different levels of ethical reasoning. *Journal of Academic Ethics* 13: 259–275.

Krathwohl, D.R. 2002. A revision of Bloom's taxonomy: An overview. *Theory Into Practice* 41 (4): 212–218.

Macer, D.R.J. 2008. *Moral Games for Teaching Bioethics*. Haifa: UNESCO.

Palys, T., and J. Lowman. 2012. Defending research confidentiality "to the extent the law allows": Lessons from the Boston College subpoenas. *Journal of Academic Ethics* 10: 271–297.

Perlman, D. 2008. Experiential ethics education: One successful model of ethics education for undergraduate nursing students in the United States. *Monash Bioethics Review* 27 (1–2): 9–32.

Potter, C. 2015. Teaching public health ethics. *Public Health Reviews* 36: 12.

Schröder-Bäck, P., P. Duncan, W. Sherlaw, C. Brall, and K. Czabanowska. 2014. Teaching seven principles for public health ethics: Towards a curriculum for a short course on ethics in public health programmes. *BMC Medical Ethics* 15: 73.

Ssebunnya, G.M. 2013. Beyond the hidden curriculum: The challenging search for authentic values in medical ethics education. *South African Journal of Bioethics and Law* 6 (2): 48–51.

Stolper, M., B. Molewijk, and G. Widdershoven. 2016. Bioethics education in clinical settings: Theory and practice of the dilemma method of moral case deliberation. *BMC Medical Ethics* 17: 45.

Taras, M. 2009. Summative assessment: The missing link for formative assessment. *Journal of Further and Higher Education* 33 (1): 57–69.

Ten Have, H. 2016. Ethics education: Global, inspiring and challenging. *International Journal of Ethics Education* 1: 1–6.

UNESCO. 2005. *Universal Declaration on Bioethics and Human Rights*. Paris: UNESCO.

———. 2008. *Bioethics Core Curriculum, Section 1: Syllabus Ethics Education Programme*. Paris: UNESCO.

———. 2011a. *Casebook on Human Dignity and Human Rights*, Bioethics Core Curriculum Casebook Series, No. 1. Paris: UNESCO.

———. 2011b. *Casebook on Benefit and Harm*, Bioethics Core Curriculum Casebook Series, No. 2. Paris: UNESCO.

Vaswani, V. 2016. Clinical ethics: Teaching. In *Encyclopedia of Global Bioethics*, ed. H.A.M.J. ten Have, 570–574. Dordrecht: Springer.

Vidal, S.M. 2016. Lifelong learning in bioethics and human rights: 10 years of the bioethics lifelong education programme in LAC. *International Journal of Ethics Education* 1: 111–125.

Wilt, K.E. 2016. Education: Simulations. In *Encyclopedia of Global Bioethics*, ed. H.A.M.J. ten Have, 1059–1067. Dordrecht: Springer.

Index

H. ten Have (ed.), *Global Education in Bioethics*, Advancing Global Bioethics 10,
https://doi.org/10.1007/978-3-319-78984-2

FSC
www.fsc.org
MIX
Papier | Fördert
gute Waldnutzung
FSC® C083411

Kolibri 360 GmbH
Ferdinand-Jühlke-Straße 7
99095 Erfurt, Deutschland
produktsicherheit@kolibri360.de